MARTIN
GARDNER
THE JINN FROM
HYPERSPACE
and Other Scribblings—Both Serious and Whimsical

59 John Glenn Drive
Amherst, New York 14228-2119

Published 2008 by Prometheus Books

Inquiries should be addressed to
Prometheus Books
59 John Glenn Drive
Amherst, New York 14228–2119
VOICE: 716–691–0133, ext. 210
FAX: 716–691–0137
WWW.PROMETHEUSBOOKS.COM

12 11 10 09 08 5 4 3 2 1

Library of Congress Cataloging-in-Publication Data

Gardner, Martin, 1914–.
The Jinn from hyperspace, and other scribblings, both serious and whimsical / by Martin Gardner.
p. cm.
ISBN 978–1–59102–565–8
1. Mathematical recreations. 2. Scientific recreations. 3. Literary recreations. 4. Literature—Miscellanea. I. Title.

QA95.G248 2007
793.74—dc22

2007038277

Printed in the United States on acid-free paper

CONTENTS

PART THREE: L. FRANK BAUM

PART FOUR: LEWIS CARROLL

To Morton Cohen
for his unrivaled contributions
to the understanding and
appreciation of Lewis
Carroll's unique genius

PREFACE

Like my earlier anthologies, this book is a haphazard collection of published (and a few unpublished) writings. The topics reflect, of course, my own peculiar interests. I have, for example, long been appalled by the media's growing obsession with pseudoscience and the paranormal. The trend is especially damaging when it concerns physical and mental health. My first two chapters deal with the thousands of lives here and abroad that have been shattered by the false memory syndrome (FMS). Both pieces were featured in *Skeptical Inquirer*, a lively periodical devoted to the debunking of bogus science—a periodical of which I had the honor of being a cofounder.

Four chapters reprint columns I contributed for several years to *Isaac Asimov's Science Fiction Magazine*. Isaac and I belonged to a mysterious New York City club called the Trapdoor Spiders. Asimov's mystery stories about the Black Widow Spiders are based on this strange group. I still miss Isaac terribly.

Occasionally I suspect I may be the only non-Catholic admirer of Gilbert Chesterton. Two chapters in this volume are from an unpublished book of essays about G. K.'s fiction. The first appeared in a liberal arts journal, the other in a magazine devoted to Chesterton. "The Great Crumpled Paper Hoax" spoofs minimal sculpture, and "So Long Old Girl" is my poetic tribute to the destroyer escort on which I served during the Second World War.

My review of Paul Davies's book gave me a chance to sound off once again in defense of the philosophy of mathematical realism. There actually are a few mathematicians convinced that all the objects and theorems of math have no reality outside human brains. Some have even likened mathematics to such cultural folkways as traffic regulations and fashions in clothes! To me such human-centered hubris is so preposterous that I find it hard to imagine how any competent mathematician could hold such a view.

Is there a common note that sounds through most of my scribblings? If so, I suppose it is a love of magic (conjuring has been a lifelong hobby). It is a Chestertonian sense of wonder, mixed with an awe close to terror, toward the existence of a fantastic universe not made by us—a monstrous mechanism containing at least one tiny globe with a surface crawling with such improbable creatures as you and me.

Martin Gardner
Norman, Oklahoma

PART ONE: SCIENCE, MATH, AND BALONEY

1.

THE FALSE MEMORY WARS

I have long been appalled and outraged by the thousands of innocent parents, relatives, and others, here and abroad, who have become victims of what is now called the false memory syndrome (FMS). Two earlier articles on the topic can be found in my *Weird Water and Fuzzy Logic* (Prometheus Books, 1996).

The following essay focuses on the role of FMS in the recent rash of arrests of Roman Catholic priests in the United States. It ran in the January/February 2006 issue of *Skeptical Inquirer*. Part 2 appeared in the following March/April issue.

In the late 1980s and early 1990s the greatest mental health scandal in North America took place. Thousands of families were cruelly ripped apart. All over the United States and Canada, previously loving adult daughters suddenly accused their fathers or other close relatives of sexually molesting them when they were young. A raft of bewildered, stricken fathers were sent to prison, some for life, by poorly informed judges and jurors. Their harsh decisions were in response to the tearful testimonies of women, most of them middle-aged, who had become convinced by a psychiatrist or social worker that they were the victims of previously forgotten pedophilia.

On what grounds were these terrible accusations made? They were (supposedly) long-repressed memories of childhood sexual abuse brought to light by self-deluded therapists using powerful suggestive techniques such as hypnotism, doses of sodium amytal (truth serum), guided imagery, dream analysis, and other dubious methods.

Of all such techniques the most worthless is hyponotism. Mesmerized patients are in a curious, little-understood state of extreme suggestibility and compliance. They will quickly pick up subtle cues about what a hypnotist wants them to say, and then say it. The notion that under hypnosis one's unconscious takes over to dredge up honest and accurate memories of a distant past event is one of the most persistent myths of psychology. There simply is no known way, short of confirming evidence, to distinguish true from false memories aroused by hypnotism or any other technique. After many sessions with a sincere but misguided therapist, false memories can become so vivid and so entrenched in a patient's mind that they will last a lifetime.

An early leader in debunking the belief that recollections of childhood trauma can be repressed for decades is the distinguished experimental psychologist Elizabeth Loftus, a professor at the University of California, at Irvine. She was awarded the prestigious $200,000 Gramemeyer Award for Psychology given annually by the University of Louisville, and also elected to the National Academy of Sciences. Her passionate book, *The Myth of Repressed Memory*, written with Katherine Katcham, has become a classic treatise on what is called the false memory syndrome (FMS). See also her article "Creating False Memories," in *Scientific American* (vol. 277, no. 3, 1997).

Another influential and tireless crusader against FMS is educator Pamela Freyd. In 1992 she established the nonprofit FMS Foundation after she and her husband, Peter, were falsely accused by their daughter of sexually molesting her when she was a child. Freyd continues to edit the foundation's bimonthly newsletter, and provides information and moral support to wrongly accused parents. By 1992 more than ten thousand distressed parents had contacted the FMS Foundation for advice on how to cope with a son or daughter's charges. Today dozens of Web sites carry on the fight against FMS.

Over the past twenty years hundreds of papers and dozens of excellent books have shed light on the FMS epidemic. In addition to Loftus's book, I reluctantly limit my list to three others: Eleanor Goldstein's *Con-*

fabulation: Creating False Memories, Destroying Families; Mark Pendergrast's *Victims of Memory: Incest Accusations and Shattered Lives*; and Frederic Crews's *The Memory Wars: Freud's Legacy in Dispute*. See also three eye-opening articles on "Recovered Memory Therapy and False Memory Syndrome," in *The Skeptic's Encyclopedia of Pseudoscience* (vol. 2), edited by Michael Shermer. Psychiatrist John Hochman's article should be read by every attorney who defends a victim of FMS. Here is his final paragraph:

> Meanwhile, there is a large FMS subculture consisting of women convinced that their "recovered memories" are accurate, therapists keeping busy doing RMT (Repressed Memory Therapy), and of authors on the "recovery" lecture and talk show circuits. In addition, there are some vocal fringes of the feminist movement that cherish RMT since it is "proof" that men are dangerous and rotten, unless proven otherwise. Skeptical challenges to RMT are met by emotional rejoinders that critics are front groups for perpetrators, and make the ridiculous analogy that "some people even say the Holocaust did not happen." RMT will eventually disappear, but it will take time.

In 2001 the FMS Foundation sent a survey questionnaire to 4,400 persons who had contacted the foundation for advice. An overwhelming number of the accusers (99 percent) were white, 93 percent were women, 86 percent were undergoing mental therapy, and 82 percent later accused their fathers of incest when they were children. Ninety-two percent said the recovery of repressed memories was the basis of their accusations.

The number of charges peaked in 1991–1992, which accounts for 34 percent of the accusations, then the rate slowly declined. By 1999–2000 the number was down to .02 percent. The decline prompted psychiatrist Paul McHugh of Johns Hopkins School of Medicine to write an optimistic article, "The End of a Delusion: The Psychiatric Memory Wars Are Over," in the *Weekly Standard* (May 26, 2003). The survey's data are analyzed and commented on in "From Refusal to Reconciliation," by McHugh, Harold Lief, Pamela Freyd, and Janet Fetkewicz, in the *Journal of Nervous and Mental Disease* (August 2004).

The authors of this valuable paper distinguish three stages of accusers:

1. *Refusers.* Those whose beliefs about past abuse are set in concrete. They refuse all contacts with anyone who does not share their convictions.
2. *Returners.* Those who return to their families but do not retract their charges or discuss them.
3. *Retractors.* Those who eventually realize that their awful memories are fabrications. They reconcile with their parents. The authors quote from a retractor's moving letter:

> I could not face the horrible thing I had done to my parents, so I had to believe the memories were true. Even though I got away from that horrible therapist, I could not go back to my entire extended family and say that I was temporarily insane and nothing had happened. It was easier to my self-esteem to pretend that I had been sexually abused by someone, and it was still my parents' fault because they should have protected me. *FMSF Newsletter*, December 1998

In the 1990s the FMS mania also blighted the lives of hundreds of preschool teachers and daycare personnel. Small children were taken by hysterical parents to trauma therapists, convinced that their children had been sexually exploited even though at first they could not recall such abuse. After many therapy sessions, repressed memories seemed to surface.

One of the most publicized cases involving preschool children concerned the Little Rascals daycare center in Edenton, North Carolina, a town decimated by the case. On the witness stand, brainwashed little rascals told wild, unbelievable tales. They "recalled" seeing the center's co-owner, Robert Kelly, murder babies. One child said "Mr. Bob" routinely shot children into outer space. Another lad told the court that Kelly had taken a group of youngsters aboard a ship surrounded by sharks. He threw a girl overboard. Asked if the sharks had eaten her, the boy replied no, he (the boy) jumped into the water and rescued her!

Robert Kelly was convicted on ninety-nine counts of first-degree sex offenses and sentenced to twelve consecutive life terms. It was the longest sentence in North Carolina history. Kelly spent six years in prison before an appeals court released him on $200,000 bond. Kelly's friends and coworkers, including his wife and the center's cook, got harsh sentences. A 1995 television documentary, *Innocence Lost*, left no doubt that the children had confabulated.

Many recent investigations have established how easily children can be led by inept therapists to imagine events that never happened. This was amusingly demonstrated by a simple experiment reported by Daniel Goleman in his article "Studies Reflect Suggestibility of Very Young Witnesses" (*New York Times*, June 11, 1993). A boy was falsely told he had been taken to a hospital to treat a finger injured by a mousetrap. In his first interview he denied this had happened. By the eleventh interview he not only recalled the event, but added many details. In fact, only extremely rarely are memories of traumatic events repressed until years later, only then emerging under suggestive therapy. On the contrary, it is far more common for victims to try vainly to forget a traumatic incident.

There are books defending the revival of long-repressed memories. By far the worst is *The Courage to Heal*, by Ellen Bass and Laura Davis. A best seller in 1988, its rhetoric persuaded tens of thousands of gullible women that their mental and behavior problems were caused by forgotten childhood sex abuse, and led them to seek validation through trauma therapy.

Another book, almost as bad, is *Secret Survivors*, by trauma therapist Sue Blume. "Incest is easily the greatest underlying reason why women seek therapy," she wrote. ". . . [i]t is not unreasonable that more than half of all women are survivors of childhood sexual trauma." Both statements are, of course, preposterous.

In 1989 Holly Ramona sought treatment for bulimia. After months of therapy by a family counselor, and later by a psychiatrist, she began to get memories of being raped by her father when she was an infant. Firm believers in Freudian symbols, Holly's two therapists convinced her that she disliked mayonnaise, soup, and melted cheese because they reminded her of her father's semen. She was unable to eat a banana unless it was sliced because it resembled her father's penis. Under oath she testified that her father had forced her to perform oral sex on the family dog!

Holly's father sued the two therapists. Lenore Terr, a psychiatrist who was an expert witness at the trial, told the jury that Holly's dislike of bananas, cucumbers, and pickles confirmed her recovered memories of being forced to perform oral sex on her father. Terr has been an "expert witness" on other similar trials. Basic Books carelessly published her shameful work, *Unchained Memories: True Stories of Traumatic Memories, Lost and Found.* Happily, a California court refused to buy Terr's Freudian speculations. Holly's father won a settlement of half a million dollars.

As the FMS plague spread, it took on ever more bizarre forms. Quack psychiatrists began regressing patients back to traumas in their mother's wombs. One therapist uncovered memories of traumas while a patient was stuck in a fallopian tube!

Those convinced that evil aliens kidnapped and tortured them with horrible experiments in hovering UFOs started to confirm their fears by repressed memory therapy. The most absurd of many books on recovered memories of flying-saucer abductions are by Temple University's historian David M. Jacobs, and two books by the late John E. Mack, a Harvard psychiatrist. Mack believed that the extraterrestrials are friendly, and come here from higher space dimensions. Harvard was unable to fire him because, like Jacobs, he had tenure. (See the interview with Mack in the *New York Times Magazine*, March 20, 1994.)

A more tragic application of FMS rested on the beliefs of countless Protestant fundamentalists that the horrors of the End Times are fast approaching. Satan, aware of the biblical prophecy that Christ will return to Earth and cast him into a lake of fire, is now on an angry rampage. He is establishing vile cults throughout the United States, Canada, and elsewhere—cults in which unspeakable rituals are performed, such as eating babies and drinking blood and urine. Dozens of shabby books about such madness have been published in spite of a thorough investigation by the FBI, which concluded that, aside from the acts of pranksters, there is no evidence that satanic cults exist here or anywhere else. In England a report by the UK Department of Health reached a similar conclusion after investigating eighty-four cases of alleged organized satanic cults.

If revived memories of cannibalizing babies are true, thousands of satanically mutilated infant bodies should be buried around the nation. Not one has been found. Why? Because, fundamentalists argue, the Devil is so powerful that he is able to obliterate all such evidence! For lurid accounts of bogus memories of satanic rituals and details about false recollections, see chapters 6 and 11 in my book *Weird Water and Fuzzy Logic* (Prometheus Books, 1996).

Among a raft of books and articles debunking the myth of satanic cults, one of the best is Lawrence Wright's "Remembering Satan" (*New Yorker*, May 17 and 24, 1993). It is reprinted in his book with the same title. Another excellent reference is sociologist Jeffrey Victor's book *Panic: The Creation of a Contemporary Legend.*

A third crazy spinoff from the false memory wars concerns New Age psychiatrists who believe in reincarnation. Under suggestive therapy, Shirley MacLaine has recalled numerous adventures experienced in her past colorful lives. A few incarnation therapists are even using hypnotism to retrieve "recollections" of events a patient will experience in *future* lives!

In 1991 Geraldo Rivera introduced three trauma survivors on his talk show. One woman said she had murdered forty babies while in a satanic cult but totally forgot about it until her memories emerged during therapy. Well-known entertainers spoke on other talk shows about their long-buried memories of pedophilia. Comedienne Roseanne Barr revealed that her parents had abused her when she was three months old! Her wild tale, vigorously denied by her dumbfounded parents, made the cover of *People* magazine. (Roseanne has since recanted.)

The memory wars are slowly subsiding, but they are still far from over. There are four reasons for the decline:

1. Reversals by enlightened appellate courts of harsh, undeserved sentences, many for life, of innocent victims of FMS.
2. The gradual education of judges, jurors, attorneys, police officers, and people in the media.
3. An increasing number of "recanters," now in the hundreds, who realize how cruelly they have been misled.
4. A growing number of large settlements of malpractice lawsuits against therapists by recanters and wrongly accused relatives.

For sensational accounts of a few such actions see chapter 11, cited earlier, of my *Weird Water* book.

2.

THE SAD CASE OF FATHER SHANLEY

In 2002, the American Catholic Church was rocked by the greatest scandal in its history. Hundreds of gay priests were defrocked for sexual abuse of altar boys and other young members of their parishes. Church superiors were bitterly attacked for efforts to conceal misconduct. Huge settlements, totaling hundreds of millions of dollars, much of it undisclosed, were paid to the victims. Payments loomed so large for a parish in Spokane, Washington, that it declared bankruptcy. More than sixty churches have closed for lack of funds.

Most of the victims had never forgotten their abuse. Moreover, there usually were full confessions by priests, or independent evidence of guilt. In February 2005, the Roman Catholic Church released some shocking figures. Over the previous three years, 5,148 American priests had faced 11,757 allegations of sexually molesting victims under eighteen. The Church's total cost of settlements, including care and counseling for both victims and abusers, legal fees, and other related expenses, was greater than $750 million. In 2003, settlement costs were $85 million in Boston's diocese alone.

In one widely publicized case, false memory syndrome played an obvious role. In 1993, Chicago's much-admired Cardinal Joseph Bernardin was sued by Steven Cook for pedophilia. His charge was based solely on memories which Cook said he had totally forgotten until they were retrieved by a trauma therapist. Later, he became convinced that his memories, which he had believed he had repressed for seventeen years, were false. He recanted. Cardinal Bernardin was too kind to sue the therapist.

The Church's sex scandal shows no sign of abating. Hundreds of lawsuits, civil and criminal, against priests now await court action. In 2002, Paul Fox Busa, an air force police officer in Colorado, then twenty-four, read an account in the *Boston Globe* of charges by the Commonwealth of Massachusetts against Paul R. Shanley. Father Shanley had been the priest of St. Jean's Church in Newton, a town near Boston, long considered the nation's most Catholic city. Busa had attended Father Shanley's church between the ages of six and twelve. The *Globe* article, Busa maintained, together with a phone conversation with a childhood friend who also claimed to have been molested by Shanley, had suddenly restored his long-repressed memories of having been raped by Shanley no less than eighty times.

So stricken was Busa by these recollections, that he said he sobbed for six hours. Nauseated, agitated, confused, he experienced convulsions and even lost faith in his religion. He considered suicide. Returning to Newton, he filed a civil suit against Cardinal Bernard Law, then head of Boston's archdiocese, for allowing Father Shanley to remain a priest even though the cardinal was aware of Shanley's compulsions.

Busa's charges led to the arrest of Shanley, then seventy-one and living in San Diego. He was handcuffed and taken back to Massachusetts, where he was jailed for seven months before being released on a $300,000 cash bond. The Church defrocked him in 2004.

Openly gay, Shanley was constantly in trouble with the Church over his sexual romps. The media dubbed him a "street priest" because of his tireless efforts to aid Boston's young drifters and drug addicts. It was widely reported that he had attended the founding meeting of the infamous NAMBLA, the North American Man/Boy Love Association. He wore his hair long, with sideburns, and favored blue jeans over the traditional black suit and white collar. One reporter described him as "part hippie, part nerd."

Shanley never denied his gay lifestyle, always insisting that his sexual escapades were with youths of consenting age, never with children. For years, Boston's archdiocese did its best to conceal Shanley's behavior by shuffling him here and there to other parishes, at the same time secretly settling three civil lawsuits against him.

Gregory Ford, the childhood pal whose phone conversation helped trigger Busa's memories, now entered the fray. He claimed to have been molested by Father Shanley when, like Busa, he had been an altar boy at St. Jean's Church. Also like Busa, Ford said he had completely forgotten these abuses until he read the *Globe*'s story about Shanley, then called the "most hated man in Massachusetts."

JoAnn Wypijewski is a New York City writer. In her beautifully composed, carefully researched article, "The Passion of Father Shanley" (*Legal Affairs*, September/October 2004), she had this to say:

> For Busa and Ford, the memories of abuse are nearly identical . . . all of the criminal activity would have occurred before the ten o'clock Sunday Mass. Until Gregory Ford came forward, none of the thousands of children who attended [the church school] while Shanley was pastor reported anything untoward. No one is on record at the time as having noticed anything unusual involving the boys and Shanley. Not their parents. Not the several women who taught the classes, including Ford's mother.

The Ford family, together with Busa and a third plaintiff, Anthony Driscoll, initiated a civil lawsuit against the Boston archdiocese. Driscoll was another of Busa's childhood friends. Like Busa and Ford, he said he could not recall abuse by Father Shanley until he read about him in the *Boston Globe*. Affidavits were filed by twenty-one other men who claimed the priest had molested them when they were children in his parish. Four of these accusers were doing time in prison for burglary. The Boston archdiocese offered a settlement of $85 million, which, incredibly, the three turned down! The lawsuit was finally settled in 2004. The *Globe* reported that Ford received $1.4 million. Busa was awarded a half million. Driscoll's take was not disclosed.

The three plaintiffs were not content with their awards. Apparently, they felt that, although Shanley had been indicted by a grand jury in 2002, he had not yet been punished for his crimes. Joined by a homeless man whose name was never revealed, the four men persuaded the Commonwealth of Massachusetts to take criminal action against the priest. Like Busa, Ford, and Driscoll, the homeless man claimed his memory of childhood abuse by Shanley did not surface until he read about the priest in a

Boston paper. He said that while on a flight to Las Vegas to gamble, he had sudden flashbacks of being molested by Shanley when he was a child in the priest's parish.

The prosecution's case began to weaken. Three of the four accusers were dropped by their attorneys. Greg Ford was the first to go. Damaging information about him slowly emerged in pretrial hearings. He had once claimed to have been raped by his father, and on another occasion, by a cousin. He had been in a mental facility seventeen times. On several occasions, he tried to kill himself. Once he threatened to burn down his house. Contradictions turned up in statements about when his memories were recovered. Prosecutors decided that Ford was a severe burden to their case.

Driscoll was the next to leave. Like the other three plaintiffs, he said his memories of Shanley's abuses were not recovered until he read about the priest in a Boston paper. Prosecuting attorneys became suspicious that he was fabricating his memories.

The nameless homeless man turned out to be an alcoholic and drug addict subject to frequent hallucinations. Twice he had attempted suicide. When he failed to show up for a hearing, and couldn't be found, the prosecution wrote him off as another liability. None of this information about the three dropouts was allowed to reach Shanley's jury.

A bizarre incident emerged in the hearings. I quote again from JoAnn Wypijewski:

> When Ford was 11, Driscoll held a sharpened pencil upright on Ford's chair as a joke. Ford sat down on it and howled in pain. . . . Doctors treated Gregory for a puncture wound to the buttock. . . . Driscoll, Busa, and Ford never forgot . . . that incident. Driscoll says he lives with guilt for hurting Ford.
>
> . . . In Ford's recovered memories, however, the pencil incident prompted Shanley to arrive on the scene and resulted not in a puncture wound but in the rape of Ford and an anal laceration. This version of the event has been repeated by Ford's parents and lawyer. It was, according to Busa, the subject of the first conversation he had with Ford, by phone in February 2002, after he heard of his friend's recovered memories.

Although false memory syndrome (FMS) played a crucial role in both the civil and criminal trials of Father Shanley, the curious fact is that the memories of all four original accusers were not based on therapy. They had allegedly been triggered by the mere reading of a newspaper! In the civil case's pretrial hearings, expert witnesses for the defense did their best to enlighten the jury about FMS. Harvard Medical School professor Harrison Pope testified that "no reputable studies have shown that traumatic memories can be repressed." Richard McNally, a Harvard experimental psychologist and author of an authoritative book, *Remembering Trauma*, told the jury, "There is just no mechanism in the mind for keeping the door shut to traumatic memory."

The moral of all this is clear. When claims are made about revived memories of pedophilia—recollections said to be repressed until uncovered decades later by suggestive therapy or by a triggering event—they should be considered fabricated fantasies unless they can be corroborated by a confession or by strong independent evidence.

"Shanley says every encounter was a willing encounter," writes JoAnn Wypijewski. "His accusers say every encounter was an abuse. Perhaps it was something in between. . . . Perhaps, given the evasions of memory, the lure of money, or self-justification, neither accused nor accusers even know the truth anymore."

On January 25, 2005, in Cambridge, Massachusetts, the criminal trial of Father Shanley got under way. The priest, then seventy-four, was accused by the Commonwealth of Massachusetts of multiple counts of rape and indecent sexual assault on a child. That child was Paul Busa. As we have seen, he remained Shanley's sole victim. The other three accusers were dropped from the case as evidence accumulated during preliminary hearings suggesting that their claims were not trustworthy.

Busa, twenty-seven, stuck by his story that Shanley had repeatedly raped and sexually molested him in other ways while he attended Shanley's St. Jean's Church from the ages of six through twelve. He claimed he had totally forgotten these abuses until he read an article about the priest in the *Boston Globe* and talked on the phone to his childhood friend Greg Ford.

Father Shanley pleaded not guilty. Gone were his dark locks, sideburns, and blue jeans. His hair, now gray-white, was closely cropped. Tall, bespectacled, wearing a neat business suit, he looked like someone's distinguished grandfather. Throughout the trial, he sat expressionless, grim-faced, never a hint of a smile. He knew that the chief prosecutor,

Lynn Rooney, had earlier sent John Geoghan, another defrocked priest, to prison, where he was soon murdered by inmates.

Was Busa actually raped eighty times by Father Shanley? Or was he the innocent victim of FMS, his memories false but vivid in his mind and honestly believed? Did he consciously fabricate his memories, basing them on abuses detailed in the *Boston Globe*'s articles and hoping for a large settlement of his civil suit? Indeed, he was given half a million dollars.

Why was Busa not content with this award, instead initiating a criminal lawsuit against the priest? It was suggested that although Shanley had been indicted by a grand jury in 2002, he had not yet been punished, and Busa wanted to see the priest suffer for his crimes. Busa's intense dislike of gays surfaced during cross-examination. He admitted that in a journal prepared for a personal injury lawyer, he had referred to Father Shanley as "that faggot."

During three days of cross-examination by Franklin Mondano, the defense attorney, Busa said that Shanley's many abuses occurred in his parish at four different places: the boys' bathroom, the rectory, the confessional, and among empty pews. Offenses included a finger penetration of his anus, repeated anal rapes, and oral sex performed by the priest. He said Father Shanley had cautioned him not to tell anyone about these incidents because no one would believe him. Most of the molestations, Busa claimed, took place after Shanley had taken him from catechism class.

Busa described a card game that resembled strip poker. It was called war, and he had forgotten the rules. Each time he lost, he had to remove an item of his clothing. When Shanley lost, he did the same. Asked what happened after both were naked, Busa replied, "I don't remember." As the years passed, Busa said, the rapes and other molestations stopped. They were replaced by erotic touching, caressing, and rubbing of his behind and crotch. After his memories of all this were revived, he became so ill and confused that he sought psychiatric help. A doctor at the air force center where Busa was stationed advised him to carry an "emotional barf bag." His body, from chest to legs, broke out in a severe rash.

Mondano brought out the damaging fact that immediately after his memories surfaced, and before he checked into the air force's hospital for psychiatric counseling, Busa telephoned a personal injury lawyer. He admitted under oath that starting in high school and lasting several years, he had a severe alcohol problem. As a young man, he had played semiprofessional baseball for several teams. Indeed, his ambition then was to rise to the major leagues. He blamed the failure of his baseball career on the unconscious influence of Father Shanley's assaults.

Busa also spoke at length of another addiction, the abuse of steroids. At times, he illegally smuggled the drug from Mexico. Busa said the main reason he took steroids was that he was then "too fat," weighing over two hundred pounds. He believed that steroids would replace his fat with muscle. When he was in high school, and his father first learned about the steroids, the father ordered his son out of the house. It was several months before Busa returned.

Busa moved many times, always needing cash and a place to bed down. There was a period, he said, when he slept on park benches and in parking lots. He never stayed long at a job. And life was difficult in the air force. He said he had been promised flight training, but he later learned that his recruiting officer had lied about this as a possibility. After his mental breakdown, the air force gave him an honorable discharge for health reasons. Busa blamed all his problems, indlucing alcohol and steroid addiction, on Father Shanley's molestations.

Oppressed by Mondano's polite but at times withering questions, at the end of two days of grilling, Busa fell apart on the stand in a fit of weeping. The next day, his wife testified that her husband had been so devastated by his recovered memories that night that he broke into sobs, got up from a sweat-soaked bed, curled into a ball on the floor, and shook uncontrollably.

Questions about the reliability or unreliability of the retrieval of long-repressed memory played a pivotal role in the trial. The prosecution's expert witness was Dr. James Chu, a clinical psychiatrist at Harvard's medical school. Chu is among those experts who dislike the term "false memory syndrome." They prefer instead to speak of "dissociative amnesia." This term reflects an unconscious ability to disconnect from one's awareness of all memories of past horrendous experiences. Truth, of course, Chu said, "is always a factor." He admitted that sometimes false memories of long-past sexual traumas can be fabricated by suggestive therapy or triggering events, but he estimated that only about 10 percent of such recovered memories are false.

How Dr. Chu arrived at 10 percent was unclear. He did say that out of nineteen of his patients who professed recovered memories of pedophilia, he was surprised to find that in seventeen cases, there was independent corroboration.

I thought Mondano's cross-questioning of Dr. Chu was unduly soft, and his final summation weak. There was no mention of the myriad cases, only a tiny fraction of which are mentioned in this article, where recovered memories of child abuse proved false. No mention of the innocent fathers, grandfathers, uncles, and daycare personnel who were sent to prison for unconfirmed pedophilia. No mention of false recollections of painful experiments on persons abducted by extraterrestrials in flying saucers. No mention of false memories of ritual horrors performed in nonexistent satanic cults. No mention of recovered memories of traumas in past life incarnations.

Two former catechism teachers testified that they had no memory of Busa ever being taken from class by Father Shanley. Four witnesses, who as children attended classes at St. Jean's parish, also testified. One woman described Busa's class as "chaotic." The rowdiest of the boys, she said, were Busa, Ford, and Driscoll. Each of the trio was often told to leave the room and stand outside the door until allowed back in. Three of the witnesses could not recall Busa ever being pulled from the class by Shanley. The fourth remembered only one such incident.

Testimonies by the former teachers and students were, of course, strong refutations of Busa's claims that his eighty rapes occurred after Shanley yanked him from class and before Shanley performed the ten o'clock mass a short time later. It is hard to believe that eighty such rapes could have taken place without a teacher or her pupils sensing that something odd was going on.

Throughout the trial, Nancy Grace, the perky host of a Court TV talk show that covered the Shanley trial, constantly expressed her outrage against psychologists and psychiatrists who maintain that memories of long-past traumas are almost never forgotten. She said she had a friend who recovered memories of child abuse, and she would stake her life on the friend's honesty. What Grace failed to understand is that patients who "recover" false memories are absolutely certain that their memories, usually created by many sessions of suggestive therapy, are genuine.

Elizabeth Loftus was her usual impressive witness for the defense. She has testified more than 250 times in recovered-memory cases and authored or coauthored some four hundred technical papers. It is easy,

Loftus argued, for false memories of pedophilia to be fabricated by suggestive therapy or by a triggering emotional event. This is not to accuse trauma therapists of deliberately trying to implant false memories. That would be criminal. Rather, there is the danger of a poorly trained therapist, believing firmly in long-forgotten abuses, unwittingly asking a leading question such as, "Were you in the bedroom when your uncle raped you?"

Loftus is well known among her peers for a famous experiment, similar to the one described in part 1 of this series, about a boy and an imaginary mousetrap injury. The adults in the experimental group were each falsely informed that as children they had been lost in a supermarket. After being told this many times, about 20 percent of the subjects not only became convinced the incident had occurred, but developed vivid memories of details.

Loftus admitted that in a 1991 paper, she stated her belief in repressed-memory therapy, but added that during the next few years, numerous research papers refuting FMS changed her mind. As for the hundreds of papers defending the recovery of long-forgotten memories, she found them "fatally flawed." Toward the close of her testimony, a comment provoked rare court laughter. Asked if she believed a jury could reach a wrong verdict, she replied, "How about O. J.?"

Although not on the witness stand, writer JoAnn Wypijewski was interviewed on Court TV. Her opening remarks:

> I just want to point out that there is real cause for skepticism when, as in this case, you had three accusers who all claimed to have suffered the exact same abuse. All forgot the abuse at exactly the same time. All remembered the abuse at the same time. All saw the same psychiatrists. All hired the same personal-injury lawyer. All received cash payments from the church. Is such a coincidence of "recovered memory" really credible?

After three of the accusers were dropped by the prosecution during preliminary hearings, Wypijewski revealed that Shanley was offered a plea bargain of two and a half years in prison if he admitted guilt. He refused. Wypijewski quotes from one of Shanley's essays: "We can say some things without question. Any sexual act with the same or opposite sex is sinful if it is rape. Or if it involves children."

I caught the end of an interview with Wypijewski on another talk show. There, she accused the media of grossly distorting and demonizing Shanley's gay lifestyle. In no way, she said, does he deserve being called a "monster." As a flagrant example of media misinformation, she cited the oft-repeated charge that Father Shanley had attended a founding meeting of NAMBLA, the North American Man/Boy Love Association. Not true. Her investigation found that not only had he not attended this meeting, but he was never a member of NAMBLA.

"I am sorry beyond telling," Father Shanley wrote in a letter that was quoted by Wypijewski in her *Legal Affairs* article, "for the wrongs of my life and for the sorrow and anguish of which I have been the occasion. How I envy those who say in their declining years: 'If I had to do all over I would not do anything differently.' For me it is the opposite: I would do many things differently. For one, I would never have become a priest and tried to wrestle with the mandatory celibacy and myriad consequences of that folly."

JoAnn Wypijewski was allowed to see Father Shanley's personal scrapbook. It contained "pictures of him as a teenage camp counselor . . . sensitive shots of sensitive boys his own age, neatly dressed, hands on hips; here a quote from Oscar Wilde, there a photograph of Rock Hudson. . . ."

Wypijewski also checked one of Shanley's diaries. From it, she excerpts the following sad, haunting entry:

> Holy Week 1972: Midnight.
>
> Somewhere on Route 78 . . . I am overwhelmed with loneliness, ashamed at my pleas to God to find a way out for me. All my prayers should be for my people for whom there is no way out. How many 16 year olds are also lonely tonight on the road, on the run? Is it really so important for me to go on? The letters say so. They warn: "If you give up so must we. You are our hope." People shouldn't put such hope in a mere man, any man. It's almost sacrilegious. If they knew the madness in me, festering below the surface, they would join the ranks of my accusers. . . . My thoughts run to that beautiful whiskey priest of Graham Green's novel, the last one left in Mexico, underground, no good, yet he cannot leave.

On February 9, 2005, the jury unanimously declared Father Shanley guilty on all counts. A week later, the judge sentenced him to twelve to fifteen years in prison. The priest was led from the courtroom in handcuffs and leg shackles. He will be eligible for parole in eight years. There will be an appeal. The jurors, ignorant of the history of false memory syndrome and its tragedies, and riveted by Paul Busa's sobbing, had no reasonable doubts.

When she testified, Busa's wife spoke directly to Father Shanley: "You are a coward who hid behind God. You are sick to your core." In a handwritten note, Busa said, "I want [Shanley] to die in prison, whether it's of natural causes or otherwise. However he dies, I hope it's slow and painful. I want him to go to hell."

If you would like timely information about the ongoing memory wars or wish to subscribe to the FMS Foundation's lively newsletter, the address is FMS Foundation, 1955 Locust Street, Philadelphia, PA 19103-5788. The foundation can also be reached at www.fmsfonline.org or 215-940-1045.

3.

PENROSE: *THE ROAD TO REALITY*

Sir Roger Penrose is among the few top physicists who share Einstein's belief—a belief he liked to say came from a little finger—that quantum mechanics is incomplete. Penrose's twistor theory of space is currently the chief rival to superstring theory—a conjecture that at the moment seems to be unraveling. His massive book (1,093 pages!) was first published in England in 2004, and later here by Knopf the same year. The book is subtitled *A Complete Guide to the Laws of the Universe*. My review ran in *New Criterion* (October 2004).

The mathematical physicist and cosmologist Roger Penrose, now professor emeritus at Oxford University, is best known to mathematicians for his discovery of Penrose tiles. These are two four-sided polygons that tile the plane only in a nonperiodic way, that is, without a fundamental region that repeats periodically like the hexagonal tiling of a bathroom floor, or the amazing tessellations of the Dutch artist M. C. Escher. To everyone's surprise, including Penrose's, his whimsical tiling turned out to underlie a previously unknown type of crystal. You can read all about this in my book *Penrose Tiles to Trapdoor Ciphers*.

Penrose's two best sellers, *The Emperor's New Mind* and its sequel, *Shadows of the Mind*, were slashing attacks on the opinions of a few artificial intelligence mavens that in just a few decades computers made with wires and switches will be able to do everything a human mind can do. Advanced computers, it was said, will someday replace the human race and colonize the cosmos! Penrose disagrees. Not until we know more about laws below the level of quantum mechanics, he argues, can computers cross that mysterious threshold separating our self-awareness from the unconscious networks of computers. Maybe the threshold will never be crossed. Computers of the sort we know how to build obviously are no more aware of what they do than a typewriter knows it is typing.

Penrose's new book, *The Road to Reality* (Knopf, 2004), is a monumental work of more than a thousand pages. Not since the publication of the redbound volumes of Richard Feynman's *Lectures on Physics* has anyone covered in such awesome detail the struggles of today's physicists to unravel the fundamental laws of our fantastic universe—to find the Holy Grail that has been called a TOE, or Theory of Everything.

Most of Penrose's masterpiece, on which he labored for eight years, is on a technical level far beyond the reach of readers unable, as Penrose warns, to handle simply fractions. The book's first half is a masterful survey of the mathematics essential for comprehending modern physics. Chapters cover hyperbolic geometry (illustrated with Escher's models of the hyperbolic plane), complex numbers (so essential in quantum mechanics), Riemann surfaces, quaternions, n-dimensional manifolds, fibre bundles, Fourier analysis, Godel's theorem, Minkowski space, Lagrangians, Hamiltonians, and other terrifying topics. Later chapters are crisp introductions to relativity, quantum theory, the big bang, black holes, time travel, and many other areas of active research. I will skip over these densely packed chapters to focus on a few aspects of Penrose's book that can be at least partly understood without mathematical fluency.

Penrose opens his mammoth treatise with a vigorous defense of Platonic realism. This is the view of almost all mathematicians and physicists. They take for granted that the objects and theorems of mathematics are timeless truths that have a strange existence independent of human minds and cultures. There is no galaxy in which two plus two is not four.

Penrose calls attention to an intricate pattern known as the Mandelbrot set. Generated on computer screens by an absurdly simple formula, this swirling pattern is so complex that successive magnifications of its parts always disclose totally unexpected properties. It is impossible, Penrose insists, to regard this mysterious pattern as something cobbled up by our minds. It existed timelessly as an abstract object, "out there," before Benoît Mandelbrot discovered it. Perhaps it exists on extraterrestrial computer printouts, perhaps in the Mind of God. Exploring it is like exploring a vast jungle.

After his sweeping survey of mathematics, Penrose takes on the daunting task of explaining quantum theory, with emphasis on its bewildering paradoxes. Consider, for example, the mind-boggling EPR paradox. Its letters are the initials of Einstein and two associates who wrote a famous paper in which they maintained that their thought experiment proves that quantum mechanics is incomplete, a view shared by Penrose.

In its simplest form, the EPR paradox imagines a quantum reaction that sends two identical particles, A and B, flying apart in opposite directions. Particle A is measured to determine if its spin is right- or left-handed. In quantum theory a particle does not have a definite spin until it is measured. Its wave function is then said to "collapse," and it acquires at random, like the heads and tails of a flipped coin, a precise handedness. Amazingly, particle B, which may be light-years from A, instantly undergoes a similar wave collapse that gives it a spin opposite the spin of A. (The conservation of angular momentum requires that A and B have opposite spins.) Now according to relativity theory no information can travel faster than light. How then does B instantly "know" the outcome of a measurement of A?

The paradox is not resolved by saying that A and B are "entangled" in a single system with a single wave function. The problem is to explain how the two particles manage to stay connected. Einstein called it a "spooky action at a distance." The EPR is only the most dramatic of many paradoxes of entanglement that have now been confirmed in laboratories. Like Einstiein, Penrose believes that such paradoxes will not be resolved until quantum mechanics is found resting on a deeper theory.

Penrose is frank in admitting that he has "prejudices" which other physicists reject. For another instance, he is not impressed by the "many-worlds interpretation" of quantum phenomena. According to this eccentric view, every time a quantum event takes place the entire universe splits into two or more parallel universes, each containing a possible outcome of the event!

Take the notorious case known as "Schrödinger's cat." Imagine a cat inside a closed box along with a Geiger counter that emits random clicks. The first click triggers a device that kills the cat. Some quantum experts, notably Eugene Wigner, believed that no quantum event is real until it is observed by a conscious mind. Until someone opens the box and looks, the poor cat is a "superposition" of two quantum states, dead and alive. In the many-worlds interpretation the cat remains alive in one world, dies in the other. This proliferation of new universes, like the forking branches of a rapidly growing tree, naturally must include duplicates of you and me!

If these billions upon billions of sprouting universes are not "real" in the same way our universe is real, but only imaginary artifacts, then the many-worlds interpretation is just another way of talking about quantum events. Yes, the talk erases some of the bizarre concepts of quantum theory, but with such an enormous violation of Occam's razor.

Quantum teleportation is another wild field of active research. Is it possible to scan an object, say an apple, and transmit to another spot its atomic structure? Will it be possible someday to teleport humans, the way they are beamed down to a planet from *Star Trek*'s spaceship? Penrose shows that such teleportations are not possible unless the original object is totally destroyed. If a person is teleported, will he be the same person after he is reconstructed? Or only a detailed copy that is a different person? Here we plunge into profound questions about human identity—questions that long bemused science fiction writers, as well as philosophers going back to John Locke and earlier.

Superstrings were believed to be inconceivably minute loops the vibrations of which generate all the basic particles. In recent years superstring theory has been absorbed into a broader conjecture called membrane theory or M-theory. Although Penrose admires M-theory's mathematical elegance, he suspects it has little relevance to the actual world. So far no way has been found to test it. Will it lead to the next great revolution in physics, as its enthusiasts hope, or will it prove to be a fad destined to go nowhere? Penrose cites numerous past conjectures that proponents thought much too beautiful not to be true, but which soon bit the dust.

The chief rival to M-theory, albeit having fewer disciples, is twistor theory. It was invented by Penrose, who, along with his colleagues, has for decades been elaborating the theory. Twistors, deriving from what are known as spinors, are abstract entities which may provide the structure of space-time. They have a permanent "chirality" (handedness). Penrose is rare among physicists in believing that the universe is fundamentally asymmetric with respect to time and chirality. There have been feeble efforts to combine parts of twistor theory and M-theory, but as things now stand, Penrose finds the two conjectures incompatible. If one is true, the other is false.

This review has given only a few fleeting glimpses into the rich abundance of Penrose's book. Not only is it admirably written, but it is also cleverly illustrated by Penrose himself. Preceding each exercise is one of three tiny icons. A smiling face tells you that the exercise is "very straightforward." A solemn face means "needs a bit of thought." And a frowning face suggests "not to be taken lightly." One drawing, which Penrose repeats twice with subtle variations, shows three spheres at the corners of a triangle. Each represents a "form of existence." One form is the Platonic realm of mathematics. Another is the physical world, and the third is the mental world. Tiny arrows show how the three worlds are clockwise connected.

The Road to Reality loops through a luxurious landscape suffused with the beauty, magic, and mystery of Being. "Why," Penrose's friend Stephen Hawking recently asked, "does the universe go to all the bother of existing?" To an atheist, G. K. Chesterton remarked in an essay on Shelley, the universe is the most exquisite masterpiece ever constructed by nobody.

For Penrose, science is a never-ending effort to penetrate the secrets of what Einstein like to call the Old One. He has no sympathy for those who think that all underlying principles of physics have now, or soon will be, discovered. (See John Horgan's book *The End of Science*.) For all we know, the universe may have infinite levels of subbasements and infinite levels of attics in the opposite direction.

Penrose opens and closes his book with two lovely parables about how intuitive insights can ignite scientific revolutions. His prologue is a tribute to Pythagoras for his discovery of irrationals, and the essential role of numbers in understanding how the Old One behaves. His postscript tells of Antea, a postdoctoral student of physics at the Albert Einstein Institute in Golm, Germany. Antea had been struggling with difficulties involving quantum gravity, black holes, and the monstrous explosion that created the universe. She had stayed up all night gazing at the stars through a large window. As dawn was about to break, she observed for the first time a rare event known as the green flash. "This experience mingled with some puzzling mathematics thoughts that had been troubling her throughout the night."

The parable's final sentence: "Then an odd thought overtook her . . ."

4.

PENROSE: *THE EMPEROR'S NEW MIND*

Will we ever be able to construct a computer, made with wires and switches, that can do everything a human mind can do? Will it be able, say, to write a great novel? If so, would such a computer be aware of itself, know what it is doing? Would it have consciousness and free will? Many artificial intelligence (AI) researchers think these things are not only possible, but may actually be achieved before the end of the century!

Sir Roger Penrose thinks this is baloney, and I agree. We belong to a group called the "mysterians" because we believe neuroscience is nowhere close to understanding consciousness; indeed, it may be that such understanding is as far beyond the capacity of our brains as calculus is beyond the understanding of a chimp. I have expressed my skepticism in earlier articles, notably in "Computers Near the Threshold?" reprinted in *The Night Is Large* (St. Martin's, 1996).

I had the privilege of writing the following piece as the foreword to Penrose's *The Emperor's New Mind* (Oxford, 1989). Penrose followed this with *Shadows of the Mind* (Oxford, 1994). Workers in AI have not taken kindly to either book.

Many great mathematicians and physicists find it difficult, if not impossible, to write a book that nonprofessionals can understand. Until this year one might have supposed that Roger Penrose, one of the world's most knowledgeable and creative mathematical physicists, belonged to such a class. Those of us who had read his nontechnical articles and lectures knew better. Even so, it came as a delightful surprise to find that Penrose had taken time off from his labours to produce a marvelous book for informed laymen. It is a book that I believe will become a classic.

Although Penrose's chapters range widely over relativity theory, quantum mechanics, and cosmology, their central concern is what philosophers call the "mind-body problem." For decades now the proponents of "strong AI" (Artificial Intelligence) have tried to persuade us that it is only a matter of a century or two (some have lowered the time to fifty years!) until electronic computers will be doing everything a human mind can do. Stimulated by science fiction read in their youth, and convinced that our minds are simply "computers made of meat" (as Marvin Minsky once put it), they take for granted that pleasure and pain, the appreciation of beauty and humour, consciousness, and free will are capacities that will emerge naturally when electronic robots become sufficiently complex in their algorithmic behavior.

Some philosophers of science (notably John Searle, whose notorious Chinese room thought experiment is discussed in depth by Penrose), strongly disagree. To them a computer is not essentially different from mechanical calculators that operate with wheels, levers, or anything that transmits signals. (One can base a computer on rolling marbles or water moving through pipes.) Because electricity travels through wires faster than other forms of energy (except light) it can twiddle symbols more rapidly than mechanical calculators, and therefore handle tasks of enormous complexity. But does an electrical computer "understand" what it is doing in a way that is superior to the "understanding" of an abacus? Computers now play grandmaster chess. Do they "understand" the game any better than a tick-tack-toe machine that a group of computer hackers once constructed with Tinkertoys?

Penrose's book is the most powerful attack yet written on strong AI. Objections have been raised in past centuries to the reductionist claim that a mind is a machine operated by known laws of physics, but Penrose's offensive is more persuasive because it draws on information not available to earlier writers. The book reveals Penrose to be more than a mathematical physicist. He is also a philosopher of first rank, unafraid to grapple with problems that contemporary philosophers tend to dismiss as meaningless.

Penrose also has the courage to affirm, contrary to a growing denial by a small group of physicists, a robust realism. Not only is the universe "out there," but mathematical truth also has its own mysterious independence and timelessness. Like Newton and Einstein, Penrose has a profound sense of humility and awe toward both the physical world and the platonic realm of pure mathematics. The distinguished number theorist Paul Erdös likes to speak of "God's book" in which all the best proofs are recorded. Mathematicians are occasionally allowed to glimpse part of a page. When a physicist or a mathematician experiences a sudden "aha" insight, Penrose believes, it is more than just something "conjured up by complicated calculation." It is mind making contact for a moment with objective truth. Could it be, he wonders, that Plato's world and the physical world (which physicists have now dissolved into mathematics) are really one and the same?

Many pages in Penrose's book are devoted to a famous fractal-like structure called the Mandelbrot set after Benoît Mandelbrot who discovered it. Although self-similar in a statistical sense as portions of it are enlarged, its infinitely convoluted pattern keeps changing in unpredictable ways. Penrose finds it incomprehensible (as do I) that anyone could suppose that this exotic structure is not as much "out there" as Mount Everest is, subject to exploration in the way a jungle is explored.

Penrose is one of an increasingly large band of physicists who think Einstein was not being stubborn or muddle-headed when he said his "little finger" told him that quantum mechanics is incomplete. To support this contention, Penrose takes you on a dazzling tour that covers such topics as complex numbers, Turing machines, complexity theory, the bewildering paradoxes of quantum mechanics, formal systems, Gödel undecidability, phase spaces, Hilbert spaces, black holes, white holes, Hawking radiation, entropy, the structure of the brain, and scores of other topics at the heart of current speculations. Are dogs and cats "conscious" of themselves? Is it possible in theory for a matter-transmission machine to translocate a person from here to there the way astronauts are beamed up and down in television's *Star Trek* series? What is the survival value that evolution found in producing consciousness? Is there a level beyond quantum mechanics in which the direction of time and the distinction between right and left are firmly embedded? Are the laws of quantum mechanics, perhaps even deeper laws, essential for the operation of a mind?

To the last two questions Penrose answers yes. His famous theory of "twistors"—abstract geometrical objects which operate in a higher-dimensional complex space that underlies space-time—are too technical for inclusion in this book. They are Penrose's efforts over two decades to probe a region deeper than the fields and particles of quantum mechanics. In his fourfold classification of theories as superb, useful, tentative, and misguided, Penrose modestly puts twistor theory in the tentative class, along with superstrings and other grand unification schemes now hotly debated.

Since 1973 Penrose has been the Rouse Ball Professor of Mathematics at Oxford University. The title is appropriate because W. W. Rouse Ball not only was a noted mathematician, he was also an amateur magician with such an ardent interest in recreational mathematics that he wrote the classic English work on this field, *Mathematical Recreations and Essays*. Penrose shares Ball's enthusiasm for play. In his youth he discovered an "impossible object" called a "tribar." (An impossible object is a drawing of a solid figure that cannot exist because it embodies self-contradictory elements.) He and his father, Lionel, a geneticist, turned the tribar into the Penrose Staircase, a structure that Maurits Escher used in two well-known lithographs: *Ascending and Descending* and *Waterfall*. One day when Penrose was lying in bed, in what he called a "fit of madness," he visualized an impossible object in four-dimensional space. It is something, he said, that a four-space creature, if it came upon it, would exclaim "My God, what's that?"

During the 1960s, when Penrose worked on cosmology with his friend Stephen Hawking, he made what is perhaps his best-known discovery. If relativity theory holds "all the way down," there must be a singularity in every black hole where the laws of physics no longer apply. Even this achievement has been eclipsed in recent years by Penrose's construction of two shapes that tile the plane, in the manner of an Escher tessellation, but which can tile it only in a nonperiodic way. (You can read about these amazing shapes in my book *Penrose Tiles to Trapdoor Ciphers*.) Penrose invented them, or rather discovered them, without any expectation they would be useful. To everybody's astonishment it turned out that three-dimensional forms of his tiles may underlie a strange new kind of matter. Studying these "quasicrystals" is now one of the most active research areas in crystallography. It is also the most dramatic instance in modern times of how playful mathematics can have unanticipated applications.

Penrose's achievements in mathematics and physics—and I have touched on only a small fraction—spring from a lifelong sense of wonder toward the mystery and beauty of being. His little finger tells him that the human mind is more than just a collection of tiny wires and switches. The Adam of his prologue and epilogue is partly a symbol of the dawn of consciousness in the slow evolution of sentient life. To me he is also Penrose—the child sitting in the third row, a distance back from the leaders of AI—who dares to suggest that the emperors of strong AI have no clothes. Many of Penrose's opinions are infused with humour, but this one is no laughing matter.

5.

TIME-REVERSED WORLDS

This reprints a chapter from my *New Ambidextrous Universe* (Dover, 2005). It is there followed by a chapter on reversed persons and particles. For related articles see the chapter on time travel in my *Time Travel and Other Mathematical Bewilderments* (W.H. Freeman, 1988) and the chapter "Does Time Ever Stop? Can the Past Be Altered?" in *Fractal Music, Hypercards, and More* (W. H. Freeman, 1992).

In my opinion, time, or its synonym change, is a terrifying mystery, perhaps the greatest of all mysteries. It is impossible to define time without introducing time into the definition. Time, said John Wheeler, is what keeps everything from happening at once. That doesn't much help. If there is a God, is God subject to time? Some theologians say no. Process theologians say yes. I don't know the answer to this question, and neither do you.

We have seen (from the book's previous chapter) how the direction of time can be defined by five arrows. Let us put aside the question of how those arrows relate to one another and consider the following question: Is it meaningful to talk about the existence of another universe, exactly like our own with respect to all basic laws, but with all five of its time arrows pointing the opposite way from ours?

The CPT (charge, parity, time) theorem suggests that such a world would be made of antimatter. There is strong evidence that antimatter (charge-reversed matter) has a handedness that is the reverse of matter. As we have learned, the violation of CP symmetry implies T (time) asymmetry. It would be aesthetically pleasing to theoreticians if a universe could exist in which C, P, and T were all reversed. As a thought experiment, let's assume that a reversal of T on the particle level is combined with a reversal of the other four time arrows. Is it possible that somewhere out there, in another space-time continuum, there is an antimatter universe in which structures not only go the other way in space, but also go the other way (in all respects) in time?

Two worlds with opposite time arrows are analogous to two worlds that are mirror images of each other. If we eliminate the role of an outside observer with his own sense of left and right, then all we can say is that each world is a mirror reversal of the other. The same is true of time-reversed worlds. In each universe intelligent beings are living "forward" in their time. To say that time in one universe is "backward" means nothing more than that events in that universe go the other way relative to events in the first universe.

This notion of two worlds with contrary arrows of time goes back to Boltzmann. There seems to be nothing logically contradictory about it, although it leads to many bizarre results. For example, no two-way communication is possible between intelligent minds in the two time-reversed worlds. To see why, suppose that some kind of communication channel is established between person A in one universe and person Z in a time-reversed world. A sends a message to Z. Z manages to decode it and send a reply back to A. From A's point of view, Z is moving into Z's past. Z can't reply because he hasn't yet got the message. From Z's point of view, any reply that he makes will arrive in A's past *before* A sent the original message! From either point of view, logical contradictions arise if the possibility of a reply is assumed. The situation is similar to those paradoxes that arise in science fiction stories when a person travels into the past and murders himself as a child.

Communication, therefore, seems to be ruled out by logic. What about observation? It is easy to see a mirror-reversed world—just look into a mirror! But seeing a time-reversed world poses difficulties. For one thing, light, instead of radiating from the other world, would be going toward it. If observation involved electromagnetic radiation, each world would be totally invisible to the other. Let us pull out all stops and suppose that someday we will discover a type of radiation that can be directed toward a time-reversed world and which will bounce back to us without interfering in any way with the other world's history. By the use of this strange radiation we can "observe" what is happening in the other world, although we cannot use it for any kind of communication. We will, of course, see the other world as moving backward in time. They in turn can use the same technique to see us moving backward in time.[1]

No one has the slightest idea of what such a radiation would be like, but there seems to be no logical contradiction that follows from assuming it to exist. Oddly enough, the assumption does not even presuppose a deterministic view of history—the view that, given the state of the universe at any one moment, the entire future of the universe is uniquely determined. When A probes the state of Z's universe, all he can observe is that universe going back into its previous states. Similarly Z, probing A's universe, sees that universe going backward. The past, as everyone agrees, is fixed for all eternity. Because neither A nor Z can probe the other world's future, both futures remain indeterminate. Seeing into the past of another universe has no more effect on the determinism-indeterminism controversy than seeing a motion picture, projected backward, of past historical events in human history. For those who participated in the events, chance and free will played their roles as the events unrolled toward an undetermined future. Paradoxes arise only if there is interaction between two worlds whose time arrows point in opposite ways. Without such interaction the differences are entirely linguistic. We would describe their events in a time-reversed language, and they would do the same if they could observe us. In each world the arrow of time would point from past to future.

Gods in higher space-times, observing two universes with opposite time arrows, are of no help in thought experiments intended to settle the controversy. Even if they see the entire history of both universes, in one blinding instant of hypertime, it does not preclude the possibility that each universe, as it unrolls in its own time, has branch points that are undetermined at each moment of branching. Indeed, this is precisely how many great theologians of all religions have reconciled the seemingly contradictory notions of free will and predestination. The ancient debate between determinists and indeterminists appears to be unaffected by the notion of time-reversed worlds.

Frank Russell Stannard, a British physicist, writing in "Symmetry of the Time Axis" (*Nature*, August 13, 1966), suggested (not too seriously) that two time-reversed worlds might occupy the same volume of space-time by interpenetrating each other but not interacting in any way, like a pair of checker players playing one game on the black squares while another pair play a different game on the red squares.[2] He called the "other" world "faustian," because Faust, in Goethe's poem, was permitted by Mephistopheles to go back in time. In Stannard's vision the faustian world is all around us, going the other way in time, forever cut off from our observation.

J. A. Lindon, my favorite writer of comic verse, was moved by Stannard's vision to compose the following poem:

NOT THAT *WAY!*

So I slipped through the doorway that said DR. STANNARD—
Oh, mental kaleidoscope! Everything swirled!
Here all that occurred seemed outlandishly mannered,
For time goes reverse *in the* faustian *world!*

I saw Mr. Crankylank backwardly biking
(But safe) through the traffic that rumbled so near;
Off home to his lunch (which had been to his liking),
The train he had missed coming in, by the rear.

I glimpsed Eddie Champer unchewing his bacon
Before turning in for his morn-to-night's rest;
His wife then unfrizzled it fit to be taken
And sold to the grocer marked BACON BACK BEST.

I bowed at the "hundreth" of old Lady Brinker,
Who died last December (though buried before);
With a chuckle this blinking fat-winking hard-drinker
Said she'd taste mother's milk in one century more!

Men showed me the prison where Bill the Bank Robber
Was serving his sentence of retrograde time;
Come seventeen years, he might doff prison clobber,
Be led out and left decommitting his crime.

I spied on Lou Cleanbody bathing. (The struggle
With my better nature defeated me, chaps!)
She soaked up the scum that emerged from the plug-hole,
Got out, and fresh water poured into the taps.

I viewed an unbombing. Debris reassembled
Itself into buildings, life came out of doom.
Smoke flashed into bombs, which flew up (as I trembled)
And hooked under planes in a tail-upward zoom.

I heard of unsabotage: works had been spannered;
The spanners flipped out, broken cogs then OK.
Here I left by the doorway that said DR. STANNARD,
And found my watch going the usual way.

But I wrote a great book on it. No one will quote it,
Though I'd learned it all thoroughly, knew it right through.
I began at the end and forgot as I wrote it,
Unscribbled the title, know no more than you!

If Stannard's vision seems fantastic, consider the following notion, which goes all the way back to Plato. Suppose that the expanding universe reaches a point at which gravitational forces halt the outward drift and the universe starts contracting. Perhaps at the extreme limit of the expansion our world will enter a space-time singularity—a point at which the equations of physics no longer apply—and then when it starts to contract, all the time arrows will spin around and point the other way. The universe, in brief, will turn into a time-reversed world of antimatter.

In recent years the Cornell astrophysicist Thomas Gold has seriously proposed just such a cosmological model,[3] but first let us see how Plato describes it in his dialogue *The Statesman*:

STRANGER: Listen, then. There is a time when God himself guides and helps to roll the world in its course; and there is a time, on the completion of a certain cycle, where he lets go, and the world being a living creature, and having originally received intelligence from its author and creator, turns about and by an inherent necessity revolves in the opposite direction.

SOCRATES: Why is that?

STRANGER: Why, because only the most divine things of all remain ever unchanged and the same, and body is not included in this class. Heaven and the universe, as we have termed them, although they have been endowed by the Creator with many glories, partake of a bodily nature, and therefore cannot be entirely free from perturbation. But their motion is, as far as possible, single and in the same place, and of the same kind; and is therefore only subject to a reversal, which is the least alteration possible. For the lord of all moving things is alone able to move of himself; and to think that he moves them at one time in one direction and at another time in another is blasphemy. Hence we must not say that the world is either self-moved always, or all made to go round by God in two opposite courses; or that two Gods, having opposite purposes, make it move round. But as I have already said (and this is the only remaining alternative) the world is guided at one time by an external power which is divine and receives fresh life and immortality from the renewing hand of the Creator, and again, when let go, moves spontaneously, being set free at such a time as to have, during infinite cycles of years, a reverse movement: this is due to its perfect balance, to its vast size, and to the fact that it turns on the smallest pivot.

SOCRATES: Your account of the world seems to be very reasonable indeed.

STRANGER: Let us now reflect and try to gather from what has been said the nature of the phenomenon which we affirmed to be the cause of all these wonders. It is this.

SOCRATES: What?

STRANGER: The reversal which takes place from time to time of the motion of the universe.

SOCRATES: How is that the cause?

STRANGER: Of all changes of the heavenly motions, we may consider this to be the greatest and most complete.

SOCRATES: I should imagine so.

STRANGER: And it may be supposed to result in the greatest changes to the human beings who are the inhabitants of the world at the time.

SOCRATES: Such changes would naturally occur.

STRANGER: And animals, as we know, survive with difficulty great and serious changes of many different kinds when they come upon them at once.

SOCRATES: Very true.

STRANGER: Hence there necessarily occurs a great destruction of them, which extends also to the life of man; few survivors of the race are left, and those who remain become the subjects of several novel and remarkable phenomena, and of one in particular, which takes place at the time when the transition is made to the cycle opposite to that in which we are now living.

SOCRATES: What is it?

STRANGER: The life of all animals first came to a standstill, and the moral nature ceased to be or look older, and was then reversed and grew young and delicate; the white locks of the aged darkened again, and the cheeks of the bearded man became smooth, and recovered their former bloom; the bodies of youths in their prime grew softer and smaller, continually by day and night returning and becoming assimilated to the nature of a newly-born child in mind as well as body; in the succeeding stage they wasted away and wholly disappeared. And the bodies of those who died by violence at that time quickly passed through the like changes, and in a few days were no more seen.

SOCRATES: Then how, Stranger, were the animals created in those days; and in what way were they begotten of one another?

STRANGER: It is evident, Socrates, that there was no such thing in the then order of nature as the procreation of animals from one another; the earth-born race, of which we hear in story, was the one which existed in those days—they rose again from the ground; and of this tradition, which is now-a-days often unduly discredited, our ancestors, who were nearest in point of time to the end of the last period and came into being at the beginning of this, are to us the heralds. And mark how consistent the sequel of the tale is; after the return of age to youth, follows the return of the dead, who are lying in the earth, to life; simultaneously with the reversal of the world the wheel of their generation has been turned back, and they are put together and rise and live in the opposite order, unless God has carried any of them away to some other lot. According to this tradition they of necessity spring from the earth and have the name of earth-born, and so the above legend clings to them.

SOCRATES: Certainly that is quite consistent with what has preceded; but tell me, was the life which you said existed in the reign of Cronos in that cycle of the world, or in this? For the change in the course of the stars and the sun must have occurred in both.

If we imagine the cycles described by Plato's stranger as repeating endlessly, we have an oscillating model of the universe that is surprisingly similar to Gold's model, as well as to the eternal recurrence doctrines of certain Eastern religions. Let us embroider it with more conjectures—first, the notion that every black hole is associated with a while hole from which all the matter and energy eaten by the black hole gush forth. The two holes are joined by an "Einstein-Rosen bridge," or what John Wheeler calls a wormhole. Perhaps the centers of quasars, and those quasarlike galaxies called Seyfert galaxies, are such white holes. If so, the final entropy death of our universe is being delayed by matter constantly being recycled through black and white holes. At the final collapse of the universe, everything will disappear into a gigantic black hole. Will this be followed by the explosion of a white hole that is the big bang of the next cycle? Are we living in a universe of what we call matter that is the left-over antimatter of the collapsing universe that preceded us?

Each backward-moving cycle in this oscillating model can be interpreted in two ways. If we assume determinism, the second cycle could simply repeat what happened in the previous cycle, but in reverse order. On the other hand, the time-reversed universe, like the faustian universe or any time-revered world situated somewhere "out there," need not be deterministic at all. It could go backward with an entirely different history. In the absence of any other universe with which it can interact, intelligent beings in the "backward" universe would find themselves moving forward in time in a perfectly normal manner.

There seems to be nothing wrong with the first interpretation except that it seems dull and pointless for history to keep repeating itself in alternate time directions. To the gods it would be like our reading *Finnegans Wake* to the end, then reading it backward to "riverrun," then forward again, and repeating this forever, or like watching a motion picture alternately projected forward and backward. But note: We need outside observers in hypertime to give a meaning to "forward" and "backward." If there are no such observers, we might just as well speak of *one* single cycle that never repeats.

The spectacle becomes less boring when the cycles are not identical in their history. Nothing in Plato's vision, or in Gold's, requires each "riverrun" to repeat its history exactly. In each cycle, intelligent beings will find the world just as we do, moving from an immutable past into an unpredictable future. The future, as William James so passionately argued, could be filled with genuine surprises that not even gods could anticipate. It would certainly make the movies more exciting for them if they didn't know how each picture ended.

Let us enlarge this vision even more. Existence contains an infinity of universes, each alternately expanding and contracting, each with genuine "futures" that do not "exist" until they happen. Intelligent creatures in any cycle of any universe deem themselves living forward in time. If there is no interaction of any sort between these worlds, it is hard to see how logical contradictions can arise to make nonsense of such a vision.

A curious thought now arises. What is to prevent radiation from the expanding half of the cycle from continuing on its way and entering into the contracting half? Paul Davies reports in his book that Wheeler has conjectured that there is a gradual "turning of the tide," when the universe slows to a halt and starts to move the other way. If so, then near the end of the expanding cycle one might begin to see a portion of this radiation coming back to us in a blurred, diffuse state. If the direction of time reverses for the contracting phase, this would be (as Davies puts it) "a search for electromagnetic microwaves from the future." At least one experiment, he tells us, has actually been made to look for such microwaves, but it failed to detect them.

Even stranger situations arise in thought experiments in which we imagine individual persons or individual particles going backward in time while the rest of the universe goes forward.

NOTES:

1. Norman Swartz, in "Is There an Ozma-Problem for Time?" (*Analysis*, 33 [January 1977]: 77–82), argues that any communication between two time-reflected worlds, or even their observation from an outside third world, would violate causal laws so radically as to be impossible.

2. There are other ways to envision two interpenetrating universes. In our chapter on superstrings we will encounter a unification theory which suggests a world of "shadow matter" that could interpenetrate the universe we know. Another possibility rests on the assumption—it is taken seriously by many mathematicians working in the field of cellular automata theory—that time is granular, a rapid succession of units that have been named *chronons*. In other words, the universe jumps from state to state like the successive frames of a motion picture, or like the changing states of pixels on a computer monitor. Action in the movies or on the computer screen appears continuous, but actually is a sequence of separate pictures with short time intervals between them. Imagine now that inside those time intervals another set of frames belongs to a different movie, but

we can see this other movie only if we run the film on a machine that displays just the second set of frames. In similar fashion, another world may be running on different time intervals, but completely inside the space of our own universe.

3. Gold's notion that entropy would go the other way in a universe that was shrinking toward the "big crunch" was even more fanciful than his recent claim that petroleum is not the product of organic life but has a chemical origin. Nevertheless, no less a person than Stephen Hawking bought it for many years. In 1986, at a conference on relativistic astrophysics in Chicago, Hawking announced a change of mind. He was convinced, he said, that if the universe went into a contracting phase, the arrow of entropy would point the same way as before.

The same can be said for the mind's psychological arrow, which Hawking regards as based on entropy. However—and he has repeated these views in his *Brief History of Time* (Bantam, 1988)—a contracting universe would soon cease to support life. See Walter Sullivan's article "Is There a Past in the Future?" (*New York Times*, December 30, 1986).

6.

THE *ARS MAGNA* OF RAMON LULL

Ever since I took a course in formal logic, at the University of Chicago, I have been fascinated by ways of solving logic problems by using diagrams. They are closely related to mechanical devices for doing the same thing. As far as I know, I was the first to write a history of these two topics, *Logic Machines and Diagrams* (McGraw Hill, 1958). The book had several reprintings by other houses, and is scheduled for a new edition by A. K. Peters. The following essay reprints the book's first chapter.

Near the city of Palma, on the island of Majorca, largest of the Balearic isles off the eastern coast of Spain, a huge saddle-shaped mountain called Mount Randa rises abruptly from a monotonously level ridge of low hills. It was this desolate mountain that Roman Lull, Spanish theologian and visionary, climbed in 1274 in search of spiritual refreshment. After many days of fasting and contemplation, so tradition has it, he experienced a divine illumination in which God revealed to him the Great Art by which he might confound infidels and establish with certainty the dogmas of his faith. According to one of many early legends describing this event, the leaves of a small lentiscus bush (a plant still flourishing in the area) became miraculously engraven with letters from the alphabets of many languages. They were the languages in which Lull's Great Art was destined to be taught.

After his illumination, Lull retired to a monastery where he completed his famous *Ars magna*, the first of about forty treatises on the working and application of his eccentric method. It was the earliest attempt in the history of formal logic to employ geometrical diagrams for the purpose of discovering nonmathematical truths, and the first attempt to use a mechanical device—a kind of primitive logic machine—to facilitate the operation of a logic system.

Throughout the remainder of Lull's colorful, quixotic life, and for centuries after his death, his Art was the center of stormy controversy. Franciscan leaders (Lull belonged to a lay order of the movement) looked kindly upon his method, but Dominicans tended to regard it as the work of a madman. Gargantua, in a letter to his son Pantagruel (Rabelais, *Gargantua and Pantagruel*, book II, chapter 8), advises him to master astronomy "but dismiss astrology and the divinatory art of Lullius as but vanity and imposture." Francis Bacon similarly ridiculed the Art in two passages of almost identical wording, one in *The Advancement of Learning* (book II), the other in *De augmentis scientiarum*, a revised and expanded version of the former book. The passage in *De augmentis* (book VI, chapter 2) reads as follows:

> And yet I must not omit to mention, that some persons, more ostentatious than learned, have laboured about a kind of method not worthy to be called a legitimate method, being rather a method of imposture, which nevertheless would no doubt be very acceptable to certain meddling wits. The object of it is to sprinkle little drops of science about, in such a manner that any sciolist may make some show and ostentation of learning. Such was the Art of Lullius: such the Typocosmy traced out by some; being nothing but a mass and heap of the terms of all arts, to the end that they who are ready with the terms may be thought to understand the arts themselves. Such collections are like a fripper's or broker's shop, that has ends of everything but nothing of worth.

Swift is thought to have had Lull's Art in mind when he described a machine invented by a professor of Laputa (*Gulliver's Travels*, part III, chapter 5). This contrivance was a twenty-foot square frame containing hundreds of small cubes linked together by wires. On each face of every cube was written a Laputan word. By turning a crank, the cubes were rotated to produce random combinations of faces. Whenever a few words happened to come together and make sense, they were copied down; then from these broken phrases erudite treatises were composed. In this manner, Swift explained, "the most ignorant person at a reasonable charge, and with a little bodily labour, may write books in philosophy, poetry, politics, law, mathematics, and theology, without the least assistance from genius or study."

On the other hand we find Giordano Bruno, the great Renaissance martyr, speaking of Lull as "omniscient and almost divine," writing fantastically elaborate treatises on the Lullian Art, and teaching it to wealthy noblemen in Venice where it had become a fashionable craze. Later we find young Leibniz fascinated by Lull's method. At the age of nineteen he wrote his *Dissertio de arte combinatoria* (Leipzig, 1666), in which he discovers in Lull's work the germ of a universal albegra by which all knowledge, including moral and metaphysical truths, can someday be brought within a single deductive system.[1] "If controversies were to arise," Leibniz later declared in an oft-quoted passage, "there would be no more need of disputation between two philosophers than between two accountants. For it would suffice to take their pencils in their hands, to sit down to their slates, and to say to each other (with a friend to witness, if they liked): Let us calculate."

These speculations of Leibniz's have led many historians to credit Lull with having foreshadowed the development of modern symbolic logic and the empiricist's dream of the "unity of science." Is such credit deserved? Or was Lull's method little more than the fatastic work of a gifted crank, as valueless as the geometric designs of medieval witchcraft? Before explaining and attempting to evaluate Lull's bizarre, now forgotten Art, it will perhaps be of interest to sketch briefly the extraordinary, almost unbelievable career of its inventor.[2]

Ramon Lull was born at Palma, probably in 1232. In his early teens he became a page in the service of King James I of Aragon and soon rose to a position of influence in the court. Although he married young and had two children, his life as a courtier was notoriously dissolute. "The beauty of women, O Lord," he recalled at the age of forty, "has been a plague and tribulation to my eyes, for because of the beauty of women have I been forgetful of Thy great goodness and the beauty of Thy works."

The story of Lull's conversion is the most dramatic of the many picturesque legends about him, and second only to Saint Augustine's as a celebrated example of a conversion following a life of indulgence. It begins with Lull's adulterous passion for a beautiful and pious married woman who failed to respond to his overtures. One day as he was riding a horse down the street he saw the lady enter church for High Mass. Lull galloped into the cathedral after her, only to be tossed out by irate worshipers. Distressed by this scene, the lady resolved to put an end to Lull's campaign. She invited him to her chamber, uncovered the bosom that he had been praising in poems written for her, and revealed a breast partially consumed by cancer. "See, Ramon," she exclaimed, "the foulness of this body that has won thy affection! How much better hadst thou done to have set thy love on Jesus Christ, of Whom thou mayest have a prize that is eternal!"

Lull retired in great shame and agitation. Shortly after this incident, while he was in his bedroom composing some amorous lyrics, he was startled by a vision of Christ hanging on the Cross. On four later occasions, so the story goes, he tried to complete the verses, and each time was interrupted by the same vision. After a night of remorse and soul searching, he hurried to morning confession as a penitent, dedicated Christian.

Lull's conversion was followed by a burning desire to win nothing less than the entire Muslim world for Christianity. It was an obsession that dominated the remainder of his life and eventually brought about his violent death. As the first necessary step in this ambitious missionary project, Lull began an intensive study of the Arabic language and theology. He purchased a Moorish slave who lived in his home for nine years, giving him instruction in the language. It is said that one day Lull struck the slave in the face after hearing him blaspheme the name of Christ. Soon thereafter the Moor retaliated by attacking Lull with a knife. Lull succeeded in disarming him and the slave was jailed while Lull pondered the type of punishment he should receive. Expecting to be put to death, the Moor hanged himself with the rope that bound him.

Before this unfortunate incident, Lull had managed to finish writing, probably in Arabic, his first book, the *Book of Contemplation*. It is a massive, dull work of several thousand pages that seeks to prove by "necessary reasons" all the major truths of Christianity. Thomas Aquinas had previously drawn a careful distinction between truths of natural theology that he believed could be established by reason, and truths of revelation that could be known only by faith. Lull found this distinction unnecessary. He believed that all the leading dogmas of Christianity, including the trinity and incarnation, could be demonstrated by irrefutable arguments, although there is evidence that he regarded "faith" as a valuable aid in understanding such proofs.

Lull had not yet discovered his Great Art, but the *Book of Contemplation* reveals his early preoccupation with a number symbolism that was characteristic of many scholars of his time. The work is divided into five books in honor of the five wounds of Christ. Forty subdivisions signify the forty days Christ spent in the wilderness. The 366 chapters are designed to be read one a day, the last chapter to be consulted only in leap years. Each chapter has ten paragraphs (the ten commandments); each paragraph has three parts (the trinity), making a total of thirty parts per chapter (the thirty pieces of silver). Angles, triangles, and circles are occasionally introduced as metaphors. Of special interest to modern logicians is Lull's practice of using letters to stand for certain words and phrases so that arguments can be condensed to almost algebraic form. For example, in chapter 335 he employs a notation of twenty-two symbols and one encounters passages such as this:

> If in Thy three properties there were no difference . . . the demonstration would five the *D* to the *H* of the *A* with the *F* and the *G* as it does with the *E*, and yet the *K* would not give significance to the *H* of any defect in the *F* or the *G*; but since diversity is shown in the demonstration that the *D* makes of the *E* and the *F* and the *G* with the *I* and the *K*, therefore the *H* has certain scientific knowledge of Thy holy and glorious Trinity.[3]

There are unmistakable hints of paranoid self-esteem in the value Lull places on his own work in the book's final chapter. It will not only prove to infidels that Christianity is the one true faith, he asserts, but it will also give the reader who follows its teaching a stronger body and mind as well as all the moral virtues. Lull expresses the wish that his book be "disseminated throughout the world," and he assures the reader that he has "neither place nor time sufficient to recount all the ways wherein this book is good and great."

These immodest sentiments are characteristic of most eccentrics who become the founders of cults, and it is not surprising to hear similar sentiments echoed by disciples of the Lullian Art in later centuries. The Old Testament was regarded by many Lullists as the work of God the Father, the New Testament, of God the Son, and the writings of Lull, of God the Holy Spirit. An oft-repeated jungle proclaimed that there had been three wise men in the world—Adam, Solomon, and Ramon:

Tres sabios hubo en el mundo,
Adán, Solomón y Raymundo.

Lull's subsequent writings are extraordinarily numerous although many of them are short and there is much repetition of material and rehashing of old arguments. Some early authorities estimated that he wrote several thousand books. Contemporary scholars consider this an exaggeration, but there is good reason to think that more than two hundred of the works attributed to him are his (the alchemical writings that bear his name are known to be spurious). Most of his books are polemical, seeking to establish Christian doctrines by means of "necessary reasons," or to combat Averroism, Judaism, and other infidel doctrines. Some are encyclopedic surveys of knowledge, such as his thirteen-hundred-page *Tree of Science* in which he finds himself forced to speak "of things in an abbreviated fashion." Many of his books are in the form of Socratic dialogues. Others are collections of terse aphorisms, such as his *Book of Proverbs*, a collection of some six thousand of them. Smaller treatises, most of which concern the application of his Great Art, are devoted to almost every subject matter with which his contemporaries were concerned—astronomy, chemistry, physics, medicine, law, psychology, mnemonics, military tactics, grammar, rhetoric, mathematics, zoology, chivalry, ethics, politics.

Very few of these polemical and pseudoscientific works have been translated from the original Catalan or Latin versions, and even in Spain they are now almost forgotten. It is as a poet and writer of allegorical romances that Lull is chiefly admired today by his countrymen. His Catalan verse, especially a collection of poems on *The Hundred Names of God*, is reported to be of high quality, and his fictional works contain such startling and imaginative conceptions that they have become an imperishable part of early Spanish literature. Chief of these allegorical books is *Blanquerna*, a kind of Catholic *Pilgrim's Progress*.[4] The protagonist, who closely resembles the author, rises through various levels of church organization until he becomes pope, only to abandon the office, amid much weeping of cardinals, to become a contemplative hermit.

The Book of the Lover and the Beloved, Lull's best-known work, is contained within *Blanquerna* as the supposed product of the hermit's pen.[5] More than any other of Lull's works, this book makes use of the phrases of human love as symbols for divine love—a practice as common in the Muslim literature prior to Lull's time as it was later to become common in the writings of Saint Theresa and other Spanish mystics. Amateur analysts who enjoy looking for erotic symbols will find *The Book of the Lover and the Beloved* a fertile field. All of Lull's passionate temperament finds an outlet here in his descriptions of the intimate relationship of the lover (himself) to his Beloved (Christ).

In Lull's other great work of fantasy, *Felix, or the Book of Marvels*, we find him describing profane love in scenes of such repulsive realism that they would shock even an admirer of Henry Miller's fiction. It is difficult not to believe that Lull's postconversion attitude toward sex had much to do with his vigorous defense of the doctrine of the immaculate conception at a time when it was opposed by the Thomists and of course long before it became church dogma.

After Lull's illumination on Mount Randa, his conviction grew steadily that in his Art he had found a powerful weapon for the conversion of the heathen. The failure of the Crusades had cast doubt on the efficacy of the sword. Lull was convinced that rational argument, aided by his method, might well become God's new means of spreading the faith. The remainder of his life was spent in restless wandering and feverish activity of a missionary and evangelical character. He gave up the large estate he had inherited from his father, distributing his possessions to the poor. His wife and children were abandoned, though he set aside funds for their welfare. He made endless pilgrimages, seeking the aid of popes and princes in the founding of schools and monasteries where his Great Art could be taught along with instruction in heathen languages. The teaching of Oriental languages to missionaries was one of Lull's dominant projects and he is justly regarded as the founder of Oriental studies in European education.

The esoteric character of his Art seems to have exerted a strong magic appeal. Schools and disciples grew so rapidly that in Spain the Lullists became as numerous as the Thomists. Lull even taught on several occasions at the great University of Paris—a signal honor for a man holding no academic degree of any kind. There is an amusing story about his attendance, when at the Sorbonne, of a class taught by Duns Scotus, then a young man fresh from triumphs at Oxford. It seems that Scotus became annoyed by the old man in his audience who persisted in making signs of disagreement with what was being said. As a rebuke, Scotus asked him the exceedingly elementary question, "What part of speech is 'Lord'?" Lull immediately replied, "The Lord is no part, but the whole," then proceeded to stand and deliver a loud and lengthy oration on the perfections of God. The story is believable because Lull always behaved as a man possessed by inspired, irrefutable truth.

On three separate occasions Lull made voyages to Africa to clash verbal swords with Saracen theologians and to preach his views in the streets of Muslim cities. On the first two visits he barely escaped with his life. Then at the age of eighty-three, his long beard snow white and his eyes burning with desire for the crown of martyrdom, he set sail once more for the northern shore of Africa. In 1315, on the streets of Bugia, he began expounding in a loud voice the errors of Muslim faith. The Arabs were understandably vexed, having twice ousted this stubborn old man from their country. He was stoned by the angry mob and apparently died on board a Genoese merchant ship to which his bruised body had been carried.[6] A legend relates that before he died he had a vision of the American continent and prophesied that a descendant (i.e., Columbus) of one of the merchants would someday discover the new world.

". . . no Spaniard since," writes Havelock Ellis (in a chapter on Lull in his *The Soul of Spain*, 1908), "has ever summed up in his own person so brilliantly all the qualities that go to the making of Spain. A lover, a soldier, something of a heretic, much of a saint, such has ever been the typical Spaniard." Lull's relics now rest in the chapel of the church of San Francisco, at Palma, where they are venerated as those of a saint, in spite of the fact that Lull has never been canonized.

In turning now to an examination of the Great Art itself,[7] it is impossible, perhaps, to avoid a strong sense of anticlimax. One wishes it were otherwise. It would be pleasant indeed to discover that Lull's method had for centuries been unjustly maligned and that by going directly to the master's own expositions one might come upon something of value that deserves rescue from the oblivion into which it has settled. Medieval scholars themselves sometimes voice such hopes. "We have also excluded the work of Raymond Lull," writes Philotheus Boehner in the introduction to his *Medieval Logic*, 1952, "since we have to confess we are not sufficiently familiar with his peculiar logic to deal with it adequately, though we suspect that it is much better than the usual evaluation by historians would lead us to believe." Is this suspicion justified? Or shall we conclude with Etienne Gilson (*History of Christian Philosophy in the Middle Ages*, 1955) that when we today try to use Lull's tables "we come up against the worst difficulties, and one cannot help wondering whether Lull himself was ever able to use them"?

Essentially, Lull's method was as follows. In every branch of knowledge, he believed, there are a small number of simple basic principles or categories that must be assumed without question. By exhausting all possible combinations of these categories we are able to explore all the knowledge that can be understood by our finite minds. To construct tables of possible combinations we call upon the aid of both diagrams and rotating circles. For example, we can list two sets of categories in two vertical columns (figure 1), then exhaust all combinations simply by drawing connecting lines as shown. Or we can arrange a set of terms in a circle (figure 2), draw connecting lines as indicated, then by reading around the circle we quickly obtain a table of two-term permutations.

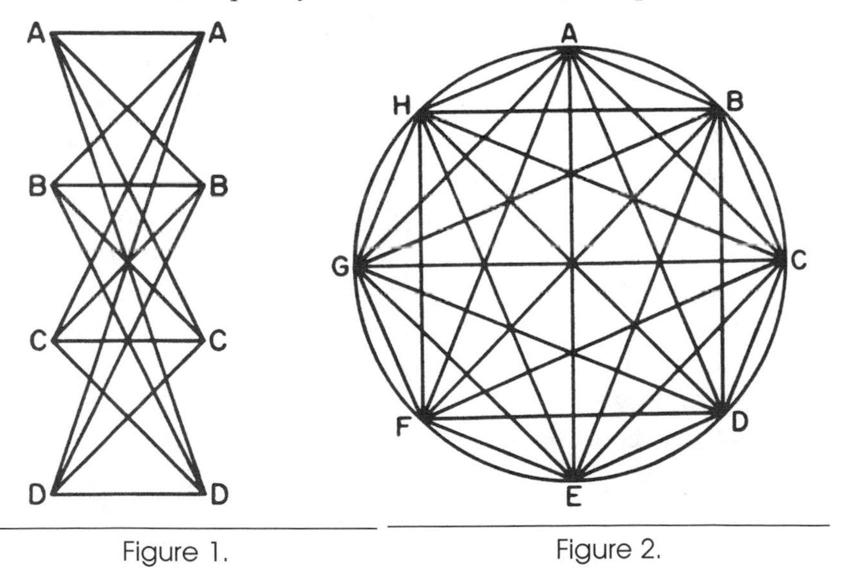

Figure 1.

Figure 2.

A third method, and the one in which Lull took the greatest pride, is to place two or more sets of terms on concentric circles as shown in figure 3. By rotating the inner circle we easily obtain a table of combinations. If

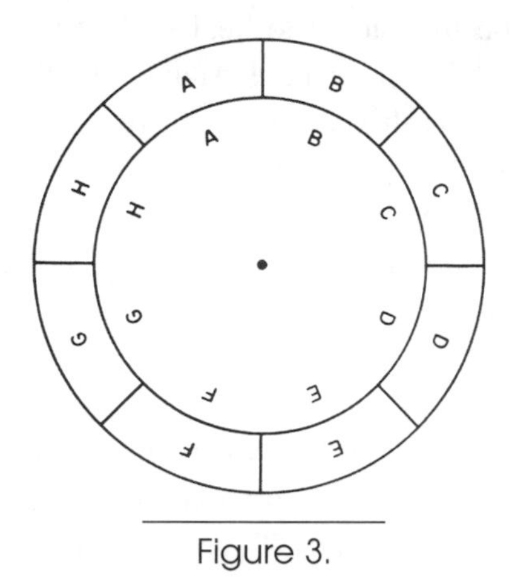

Figure 3.

there are many sets of terms that we wish to combine, this mechanical method is much more efficient than the others. In Lull's time these circles were made of parchment or metal and painted vivid colors to distinguish different subdivisions of terms. There is no doubt that the use of such strange, multicolored devices threw an impressive aura of mystery around Lull's teachings that greatly intrigued men of little learning, anxious to find a shortcut method of mastering the intricacies of scholasticism. We find a similar appeal today in the "structural differential" invented by Count Alfred Korzybski to illustrate principles of general semantics. Perhaps there is even a touch of the same awe in the reverence with which some philosophers view symbolic logic as a tool of philosophical analysis.

Before going into the more complicated aspects of Lull's method, let us give one or two concrete illustrations of how Lull used his circles. The first of his seven basic "figures" is called *A*. The letter "*A*," representing God, is placed in the center of a circle. Around the circumference, inside sixteen compartments (or "camerae" as Lull called them), we now place the sixteen letters from *B* through *R* (omitting *J*, which had no existence in the Latin of the time). These letters stand for sixteen divine attributes—*B* for goodness (*bonitas*), *C* for greatness (*magnitudo*), *D* for eternity (*eternitas*), and so on. By drawing connecting lines (figure 4) we obtain 240 two-term permutations of the letters, or 120 different combinations that can be arranged in a neat triangular table as shown at right.

BC	*BD*	*BE*	*BF*	*BG*	*BH*	*BI*	*BK*	*BL*	*BM*	*BN*	*BO*	*BP*	*BQ*
	CD	*CE*	*CF*	*CG*	*CH*	*CI*	*CK*	*CL*	*CM*	*CN*	*CO*	*CP*	*CQ*
		DE	*DF*	*DG*	*DH*	*DI*	*DK*	*DL*	*DM*	*DN*	*DO*	*DP*	*DQ*
			EF	*EG*	*EH*	*EI*	*EK*	*EL*	*EM*	*EN*	*EO*	*EP*	*EQ*
				FG	*FH*	*FI*	*FK*	*FL*	*FM*	*FN*	*FO*	*FP*	*FQ*
					GH	*GI*	*GK*	*GL*	*GM*	*GN*	*GO*	*GP*	*GQ*
						HI	*HK*	*HL*	*HM*	*HN*	*HO*	*HP*	*HQ*
							IK	*IL*	*IM*	*IN*	*IO*	*IP*	*IQ*
								KL	*KM*	*KN*	*KO*	*KP*	*KQ*
									LM	*LN*	*LO*	*LP*	*LQ*
										MN	*MO*	*MP*	*MQ*
											NO	*NP*	*NQ*
												OP	*OQ*
													PQ

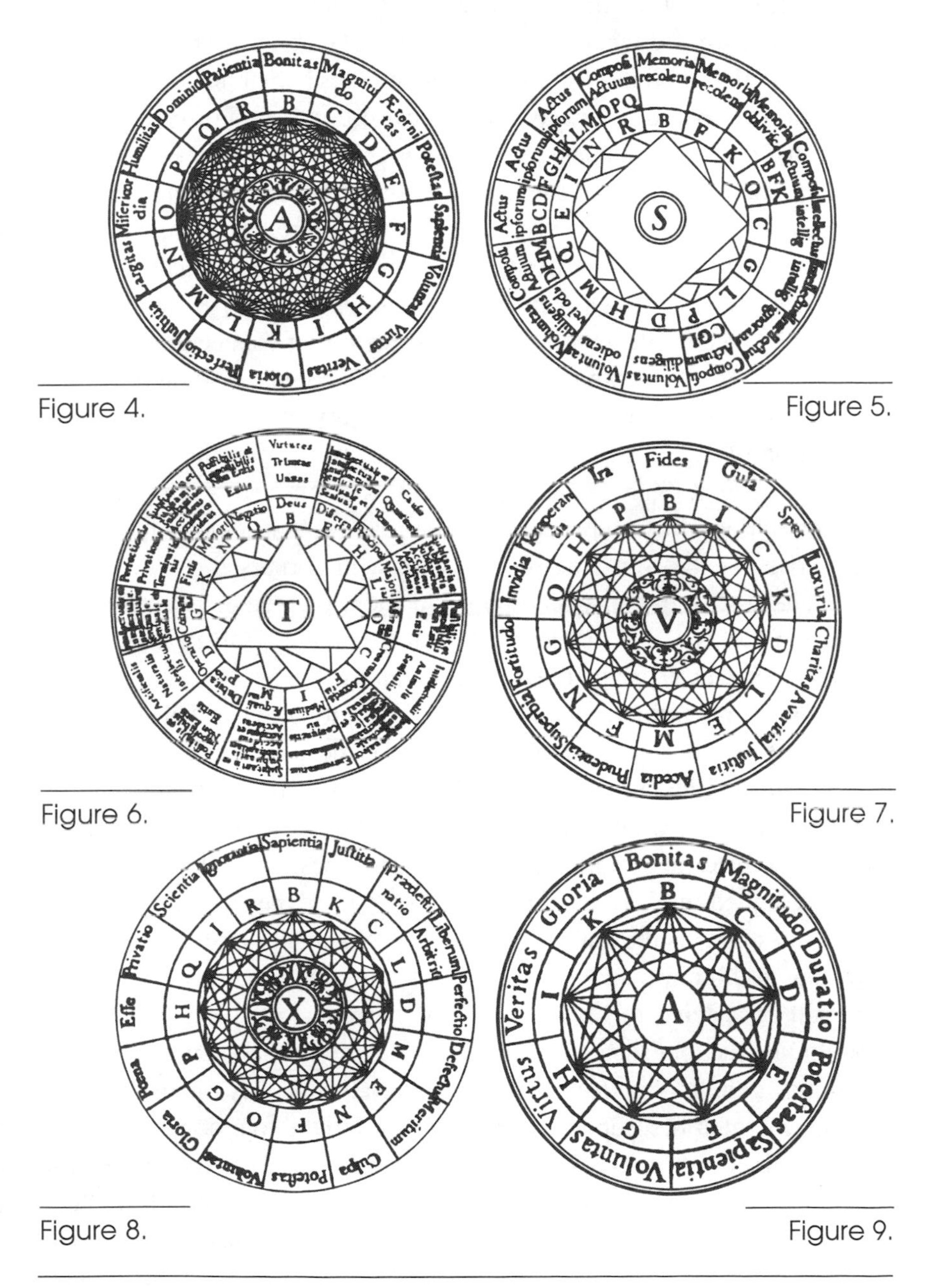

Figure 4.

Figure 5.

Figure 6.

Figure 7.

Figure 8.

Figure 9.

From the *Enciclopedia universal ilustrada*, Barcelona, 1923.

Each of the above combinations tells us an additional truth about God. Thus we learn that His goodness is great (*BC*) and also eternal (*BD*), or to take reverse forms of the same pairs of letters, His greatness is good (*CB*) and likewise His eternity (*DB*). Reflecting on these combinations will lead us toward the solution of many theological difficulties. For example, we realize that predestination and free will must be combined in some mysterious way beyond our ken; for God is both infinitely wise and infinitely just; therefore He must know every detail of the future, yet at the same time be incapable of withholding from any sinner the privilege of choosing the way of salvation. Lull considered this a demonstration "*per aequiparantium*," or by means of equivalent relations. Instead of connecting ideas in a cause-and-effect chain, we trace them back to a common origin. Free will and predestination sprout from equally necessary attributes of God, like two twigs growing on branches attached to the trunk of a single tree.

Lull's second figure [figure 5] concerns the soul and is designated by the letter *S*. Four differently colored squares are used to represent four different states of the soul. The blue square, with corners *B*, *C*, *D*, *E*, is a normal, healthy soul. The letters signify memory that remembers (*B*), intellect that knows (*C*), will that loves (*D*), and the union of these three faculties (*E*). The black square (*FGHI*) is the condition that results when the will hates in a normal fashion as, for example, when it hates sin. This faculty is symbolized by the letter *H*. *F* and *G* stand for the same faculties as *B* and *C*, and *I* for the union of *F*, *G*, and *H*. The red square (*KLMN*) denotes a condition of soul in which the memory forgets (*K*), the mind is ignorant (*L*), and the will hates in an abnormal fashion (*M*). These three degenerate faculties are united in *N*. The green square (*OPQR*) is the square of ambivalence or doubt. *R* is the union of a memory that retains and forgets (*O*), a mind that both knows and is ignorant (*P*), and a will that loves and hates (*Q*). Lull considered this last state the unhealthiest of the four. We now superimpose the four squares (figure 5) in such a way that their colored corners form a circle of sixteen letters. This arrangement is more ingenious than one might at first suppose. For in addition to the four corner letters *E*, *I*, *N*, *R*, which are unions of the other three corners of their respective squares, we also find that the faculties *O*, *P*, and *Q* are unions of the three faculties that precede them as we move clockwise around the figure. The circle of sixteen letters can now be rotated within a ring of compartments containing the same faculties to obtain 136 combinations of faculties.

It would be impossible and profitless to describe all of Lull's scores of other figures, but perhaps we can convey some notion of their complexity. His third figure, *T*, concerns relations between things. Five equilateral triangles of five different colors are superimposed to form a circle of fifteen letters, one letter at each vertex of a triangle (figure 6). As in the previous figure, the letters are in compartments that bear the same color as the polygon for which they mark the vertices. The meanings of the letters are: God, creature, and operation (blue triangle); difference, similarity, and contrariety (green); beginning, middle, and end (red); majority, equality, and minority (yellow); affirmation, negation, and doubt (black). Rotating this circle within a ring bearing the same fifteen basic ideas (broken down into additional elements) gives us 120 combinations, excluding pairs of the same term (*BB*, *CC*, etc.) We are thus able to explore such topics as the beginning and end of God, differences and similarities of animals, and so on. Lull later found it necessary to add a second figure *T*, formed of five tinted triangles whose vertices stand for such concepts as before, after, superior, inferior, universal, particular, and so on. This likewise rotated within a ring to produce 120 combinations. Finally, Lull combined the two sets of concepts to make thirty in all. By placing them on two circles he obtained 465 different combinations.

Lull's fourth figure, which he called *V*, deals with the seven virtues and the seven deadly sins. The fourteen categories are arranged alternately around a circle in red (sinful) and blue (virtuous) compartments (figure 7). Drawing connecting lines, or rotating the circle within a similarly labeled ring, calls our attention to such questions as when it might be prudent to become angry, when lust is the result of slothfulness, and similar matters. Lull's figure *X* employs eight pairs of traditionally opposed terms, such as being (*esse*) and privation (*privatio*), arranged in alternate blue and green compartments (figure 8). Figures *Y* and *Z* are undivided circles signifying, respectively, truth and falsehood. Lull used these letters occasionally in connection with other figures to denote the truth or falsehood of certain combinations of terms.

This by no means exhausts Lull's use of rotating wheels. Hardly a science or subject matter escapes his analysis by this method. He even produced a book on how preachers could use his Art to discover new topics for sermons, supplying the reader with 100 sample sermons produced by his spinning wheels! In every case the technique is the same: find the basic elements, then combine them mechanically with themselves or with the elements of other figures. Dozens of his books deal with applications of the Art, introducing endless small variations of terminology and symbols. Some of these works are introductions to more comprehensive treatises. Some are brief, popular versions for less intellectual readers who find it hard to comprehend the more involved figures. For example, the categories of certain basic figures are reduced from sixteen to nine (see figure 9). These simpler ninefold circles are the ones encountered in the writing of Bruno, Kircher, and other Renaissance Lullists, in Hegel's description of the Art (*Lectures on the History of Philosophy*, vol. 3), and in most modern histories of thought that find space for Lull's method. Two of Lull's treatises on his Art are written entirely in Catalan verse.

One of Lull's ninefold circles is concerned with objects of knowledge—God, angel, heaven, man, the imagination, the sensitive, the negative, the elementary, and the instrumental. Another asks the nine questions—whether? what? whence? why? how great? of what kind? when? where? and how? Many of Lull's books devote considerable space to questions suggested by these and similar circles. *The Book of the Ascent and Descent of the Intellect*, using a twelvefold and a fivefold circle in application to eight categories (stone, flame, plant, animal, man, heaven, angel, God) considers such scientific posers as: Where does the flame go when a candle is put out? Why does rue strengthen the eyes and onions weaken them? Where does the cold go when a stone is warmed?

In another interesting work Lull uses his Art to explain to a hermit the meaning of some of the *Sentences* of Peter Lombard. The book takes up such typical medieval problems as: Could Adam and Eve have cohabited before they ate their first food? If a child is slain in the womb of a martyred mother, will it be saved by a baptism of blood? How do angels speak to each other? How do angels pass from one place to another in an instant of time? Can God make matter without form? Can He damn Peter and save Judas? Can a fallen angel repent? In one book, the *Tree of Science*, over four thousand such questions are raised! Sometimes Lull gives the combination of terms in which the answer may be found, together with a fully reasoned commentary. Sometimes he merely indicates the figures to be used, letting the reader find the right combinations for himself. At other times he leaves the question entirely unanswered.

The number of concentric circles to be used in the same figure varies from time to time—two or three being the most common. The method reaches it climax in a varicolored metal device called the *figura universalis*, which has no less than fourteen concentric circles! The mind reels at the number and complexity of topics that con be explored by this fantastic instrument.

Before passing on to an evaluation of Lull's method, it should be mentioned that he also frequently employed the diagrammatic device of the tree to indicate subdivisions of genera and species. For Lull it was both an illustrative and a mnemonic device. His *Principles of Medicine*, for example, pictures his subject matter as a tree with four roots (the four humors) and two trunks (ancient and modern medicine). The trunks branch off into various boughs on which flowers bloom, each flower having a symbolic meaning (air, exercise, food, sleep, etc.). Colored triangles, squares, and other Lullian figures also are attached to the branches.

None of Lull's scientific writings, least of all his medical works, added to the scientific knowledge of his time. In such respects he was neither ahead nor behind his contemporaries. Alchemy and geomancy he rejected as worthless. Necromancy, or the art of communicating with the dead, he accepted in a sense common in his day and still surviving in the attitude of many orthodox churchmen; miraculous results are not denied, but they are regarded as demonic in origin. Lull even used the success of necromancers as a kind of proof of the existence of God. The fallen angels could not exist, he argued, if God had not created them.

There is no doubt about Lull's complete acceptance of astrology. His so-called astronomical writings actually are astrological, showing how his circles can be used to reveal various favorable and unfavorable combinations of planets within the signs of the zodiac. In one of his books he applies astrology to medicine. By means of the Art he obtains combinations of the four elements (earth, air, fire, water) and the four properties (hot, cold, moist, dry). These are then combined in various ways with the signs of the zodiac to answer medical questions concerning diet, evacuation, preparation of medicines, fevers, color of urine, and so on.

There is no indication that Ramon Lull, the Doctor Illuminatus as he was later called, ever seriously doubted that his Art was the product of divine illumination. But one remarkable poem, the *Desconort* ("Disconsolateness"), suggests that at times he may have been tormented by the thought that possibly his Art was worthless. The poem is ingeniously constructed of sixty-nine stanzas, each consisting of twelve lines that end in the same rhyme. It opens with Lull's bitter reflections on his failure for the past thirty years to achieve any of his missionary projects. Seeking consolation in the woods, he comes upon the inevitable hermit and pours out to him the nature of his sorrows. He is a lonely, neglected man. People laugh at him and call him a fool. His great Art is ridiculed and ignored. Instead of sympathizing, the hermit tries to prove to Ramon that he deserves this ridicule. If his books on the Art are read by men "as fast as a cat that runs through burning coals," perhaps this is because the dogmas of the church cannot be demonstrated by reason. If they could be, then what merit would there be in believing them? In addition, the hermit argues, if Lull's method is so valuable, how is it that the ancient philosophers never thought of it? And if it truly comes from God, what reason has he to fear it will ever be lost?

Lull replies so eloquently to these objections that we soon find the hermit begging forgiveness for all he has said, offering to join Ramon in his labors, and even weeping because he had not learned the Art earlier in life!

Perhaps the most striking illustration of how greatly Lull valued his method is the legend of how he happened to join the third order of Franciscans. He had made all necessary arrangements for his first missionary trip to North Africa, but at the last moment, tormented by doubts and fears of imprisonment and death, he allowed the boat to sail without him. This precipitated a mental breakdown that threw him into a state of profound depression. He was carried into a Dominican church and while praying there he saw a light like a star and heard a voice speak from above: "Within this order thou shalt be saved." Lull hesitated to join the order because he knew the Dominicans had little interest in his Art whereas the Franciscans had found it of value. A second time the voice spoke from the light, this time threateningly: "And did I not tell thee that only in the order of the Preachers thou wouldst find salvation?" Lull finally decided it would be better to undergo personal damnation than risk the loss of his Art whereby others might be saved. Ignoring the vision, he joined the Franciscans.

It is clear from Lull's writings that he thought of his method as possessing many values. The diagrams and circles aid the understanding by making it easy to visualize the elements of a given argument. They have considerable mnemonic value, an aspect of his Art that appealed strongly to Lull's Renaissance admirers. They have rhetorical value, not only arousing interest by their picturesque, cabalistic character, but also aiding in the demonstration of proofs and the teaching of doctrines. It is an investigative and inventive art. When ideas are combined in all possible ways, the new combinations start the mind thinking along novel channels and one is led to discover fresh truths and arguments, or to make new inventions. Finally, the Art possesses a kind of deductive power.

Lull did not, however, regard his method as a substitute for the formal logic of Aristotle and the schoolmen. He was thoroughly familiar with traditional logic and his writings even include the popular medieval diagrams of immediate inference and the various syllogistic figures and moods. He certainly did not think that the mere juxtaposition of terms provided in themselves a proof by "necessary reasons." He did think, however, that by the mechanical combination of terms one could discover the necessary building blocks out of which valid arguments could then be constructed. Like his colleagues among the schoolmen, he was convinced that each branch of knowledge rested on a relatively few, self-evident principles which formed the structure of all knowledge in the same way that geometrical theorems were formed out of basic axioms. It was natural for him to suppose that by exhausting the combinations of such principles one might thereby explore all possible structures of truth and so obtain universal knowledge.

There is a sense, of course, in which Lull's method of exploration does possess a formal deductive character. If we wish to exhaust the possible combinations of given sets of terms, then Lull's method obviously will do this for us in an irrefutable way. Considered mathematically, the technique is sound, though even in its day it was essentially trivial. Tabulating combinations of terms was certainly a familiar process to mathematicians as far back as the Greeks, and it would be surprising indeed if no one before Lull had thought of using movable circles as a device for obtaining such tables. Lull's mistake, in large part a product of the philosophic temper of his age, was to suppose that his combinatorial method had useful application to subject matters where today we see clearly that it does not apply. Not only is there a distressing lack of "analytic" structure in areas of knowledge outside of logic and mathematics, there is not even agreement on what to regard as the most primitive, "self-evident" principles in any given subject matter. Lull naturally chose for his categories those that were implicit in the dogmas and opinions he wished to establish. The result, as Chesterton might have said, was that Lull's circles led him in most cases into proofs that were circular. Other schoolmen were of course often guilty of question begging, but it was Lull's peculiar distinction to base this type of reasoning on such an artificial, mechanical technique that it amounted virtually to a satire of scholasticism, a sort of hilarious caricature of medieval argumentation.

We have mentioned earlier that it was Leibniz who first saw in Lull's method the possibility of applying it to formal logic.[8] For example, in his *Dissertio de arte combinatoria* Leibniz constructs an exhaustive table of all possible combinations of premises and conclusions in the traditional syllogism. The false syllogisms are then eliminated, leaving no doubt as to the number of valid ones, though of course revealing nothing that was not perfectly familiar to Aristotle. A somewhat similar technique of elimination was used by Jevons in his "logical alphabet" and his logic machine, and is used today in the construction of matrix tables for problems in symbolic logic. Like Lull, however, Leibniz failed to see how restricted was the application of such a technique, and his vision of reducing all knowledge to composite terms built up out of simple elements and capable of being manipulated like mathematical symbols is certainly as wildly visionary as Lull's similar dream. It is only in the dimmest sense that Leibniz can be said to anticipate modern symbolic logic. In Lull's case the anticipation is so remote that it scarcely deserves mention.

Still, there is something to be said for certain limited applications of Lull's circles, though it must be confessed that the applications are to subject matters which Lull would have considered frivolous. For example, parents seeking a first and middle name for a newborn baby might find it useful to write all acceptable first names in one circle and acceptable middle names on a larger circle, then rotate the inner circle to explore the possible combinations. Ancient coding and decoding devices for secret ciphers make use of Lullian-type wheels. Artists and decorators sometimes employ color wheels for exploring color combinations. Anagram puzzles often can be solved quickly by using Lullian circles to permute the required letters. A cardboard toy for children consists of a rotating circle with animal pictures around the circumference, half of each animal on the circle and half on the sheet to which the wheel is fastened. Turning the circle produces amusing combinations—a giraffe's head on the body of a hippopotamus, and so on. One thinks also of Sam Loyd's famous "Get off the earth" paradox. Renan once described Lull's circles as "magic," but in turning Loyd's wheel the picture of an entire Chinese warrior is made to vanish before your very eyes.[9] It is amusing to imagine how Lull would have analyzed Loyd's paradox, for his aptitude for mathematical thinking was not very high.

Even closer to the spirit of Lull's method is a device that was sold to fiction writers many years ago and titled, if I remember correctly, the "Plot Genii." By turning concentric circles one could obtain different combinations of plot elements. (One suspects that Aldous Huxley constructed his early novels with the aid of wheels bearing different neurotic types. He simply spun the circles until he found an amusing and explosive combination of house guests.) Mention also should be made of the book called *Plotto*, privately published in Battle Creek, Michigan, 1928, by William Wallace Cook, a prolific writer of potboilers. Although *Plotto* did not use spinning wheels, it was essentially Lullian in its technique of combining plot elements, and apparently there were many writers willing to pay the seventy-five-dollar price originally asked by the author.

In current philosophy one occasionally comes upon notions for which a Lullian device might be appropriate. For instance, Charles Morris tells us that a given sign (e.g., a word) can be analyzed in terms of three kinds of meaning: syntactic, semantic, and pragmatic. Each meaning in turn has a syntactic, semantic, and pragmatic meaning, and this threefold analysis can be carried on indefinitely. To dramatize this dialectical process one might use a series of rotating circles, each bearing the words "syntactic," "semantic," and "pragmatic," with the letter *S* in the center of the inner wheel to signify the sign being analyzed.

In science there also are rare occasions when a Lullian technique might prove useful. The tree diagram is certainly a convenient way to picture evolution. The periodic table can be considered a kind of Lullian chart that exhausts all permissible combinations of certain primitive principles and by means of which chemists have been able to predict the properties of elements before they were discovered. Lull's crude anticipation was a circle bearing the four traditional elements and rotated within a ring similarly labeled.

There may even be times when an inventor or researcher might find movable circles an aid. Experimental situations often call for a testing of all possible combinations of a limited number of substances or techniques. What is invention, after all, except the knack of finding new and useful combinations of old principles? When Thomas Edison systematically tested almost every available substance as a filament for his lightbulb, he was following a process that Lull would probably have considered an extension of his method. One American scientist, an acoustical engineer and semiprofessional magician, Dariel Fitzkee, actually published in 1944 a book called *The Trick Brain* in which he explains a technique for combining ideas in Lullian fashion for the purpose of inventing new magic tricks.

If the reader will take the trouble to construct some Lullian circles related to a subject matter of special interest to himself, and play with them for a while, he will find it an effective way of getting close to Lull's mind. There is an undeniable fascination in twisting the wheels and letting the mind dwell on the strange combinations that turn up. Something of the mood of medieval Lullism begins to pervade the room and one comprehends for the first time why the Lullian cult persisted for so many centuries.

For persist it did.[10] Fifty years after Lull's death it was strong enough to provoke a vigorous campaign against Lullism, led by Dominican inquisitors. They succeeded in having Lull condemned as a heretic by a papal bull, though later church officials decided that the bull had been a forgery. Lullist schools, supported chiefly by Franciscans, flourished throughout the late Middle Ages and Renaissance, mostly in Spain but also in other parts of Europe. We have already cited Bruno's intense interest in the Art. The great ex-Dominican considered Lull's method divinely inspired though badly applied. For example, he thought Lull mad to suppose that such truths of faith as the incarnation and trinity could be established by necessary reasons. Bruno's first and last published works, as well as many in between, were devoted to correcting and improving the method, notably *The Compendious Building and Completion of the Lulliian Art.*

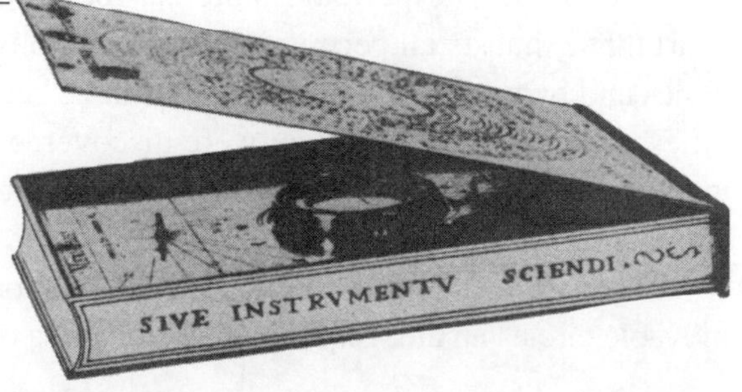

Figure 10: Sixteenth-century portable sundial engraved with Lullian figures. From *Acta Archaeologica* (1925), Wiley-Blackwell Publishing Ltd. Reprinted with permission.

In 1923 the British Museum acquired a portable sundial and compass made in Rome in 1593 in the form of a book (figure 10). On the front and back of the two gilt Copper "covers" are engraved the Lullian circles shown in figures 11 to 14. For an explanation of these circles the reader is referred to O. M. Dalton's article, "A Portable Dial in the Form of a Book, with Figures Derived from Ramon Lul," *Archaeologica*, vol. 74, second series, Oxford, 1925, pp. 89-102.

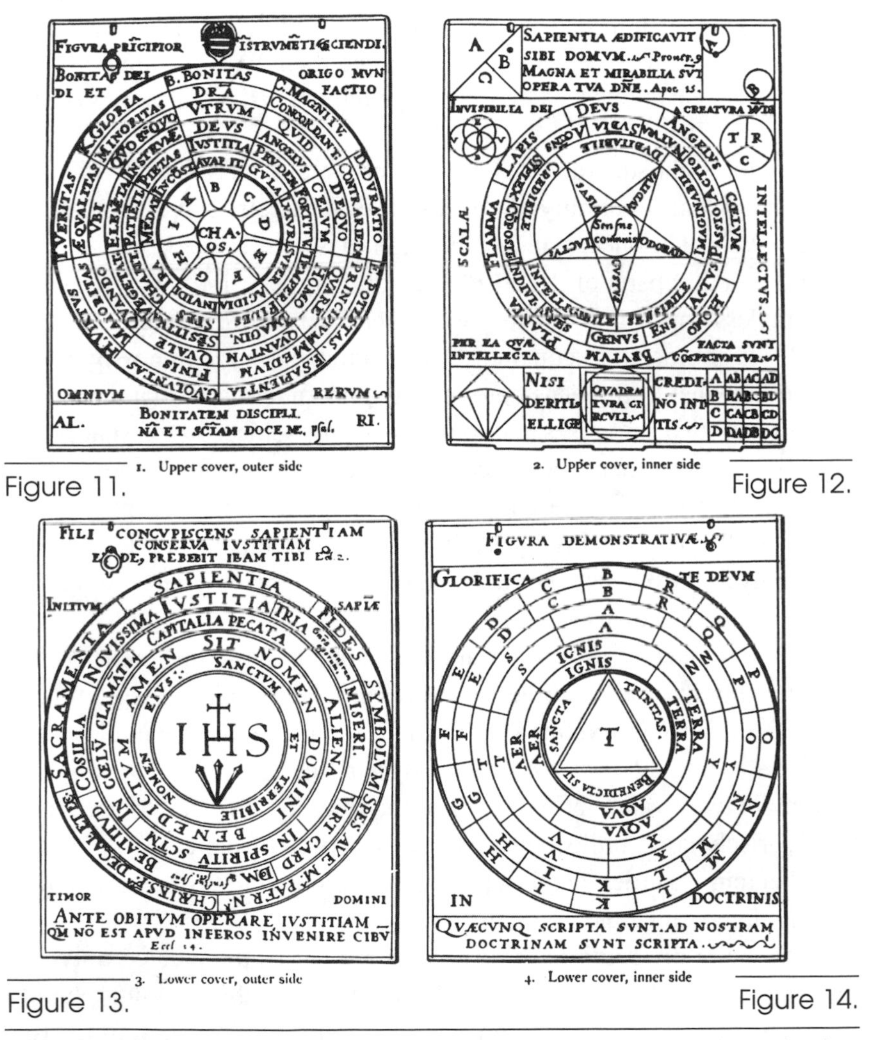

1. Upper cover, outer side

Figure 11.

2. Upper cover, inner side

Figure 12.

3. Lower cover, outer side

Figure 13.

4. Lower cover, inner side

Figure 14.

Circles used by Renaissance Lullists. From *Acta Archaeologica* (1925), Wiley-Blackwell Publishing Ltd. Reprinted with permission.

The seven smaller diagrams in figure 12 are all from Lull's writings[11] and perhaps worth a few comments. The square in the upper left corner is designed to show how the mind can conceive of geometrical truths not apparent to the senses. A diagonal divides the square into two large triangles, one of which is subdivided to make the smaller triangles *B* and *C*. Each triangle contains three angles; so that our senses immediately perceive nine angles in all. However, we can easily imagine the large triangle to be subdivided also, making four small triangles or twelve angles in all. The three additional angles exist "potentially" in triangle *A*. We do not see them with our eyes, but we can see them with our imagination. In this way our intellect, aided by imagination, arrives at new geometrical truths.

The top right square is designed to prove that there is only one universe rather than a plurality of worlds. The two circles represent two universes. We see at once that certain parts of *A* and *B* are nearer to each other than other parts of *A* and *B*. But, Lull argues, "far" and "near" are meaningless concepts if nothing whatever exists in the space between *A* and *B*. We are forced to conclude that two universes are impossible.

I think what Lull means here, put in modern terms, is that we cannot conceive of two universes without supposing some sort of space-time relations between them, but once we relate them, we bring them into a common manifold; so we can no longer regard them as separate universes. Lull qualifies this by saying that his argument applies only to actual physical existence, not to higher realms of being which God could create at will, since His power is infinite.

The four intersecting circles are interesting because they anticipate in a vague way the use of circles to represent classes in the diagrammatic methods of Euler and Venn. The four letters which label the circles stand for *Esse* (being), *Unum* (the one), *Verum* (the true), and *Bonum* (the good). *Unum*, *verum*, and *bonum* are the traditional three "transcendentales" of scholastic philosophy. The overlapping of the circles indicates that the four qualities are inseparable. Nothing can exist without possessing unity, truth, and goodness.

The circle divided into three sectors represents the created universe, but I am not sure of the meaning of the letters which apparently signify the parts. The lower left square illustrates a practical problem in navigation. It involves a ship sailing east, but forced to travel in a strong north wind. The lower right square is clearly a Lullian table displaying the twelve permutations of *ABCD* taken two letters at a time.

The remaining diagram, at the middle of the bottom, is a primitive method of squaring the circle and one fairly common in medieval pseudomathematical works. We first inscribe a square and circumscribe a square; then we draw a third square midway between the other two. This third square, Lull mistakenly asserts, has a perimeter equal to the circumference of the circle as well as an area equal to the circle's area. Lull's discussion of this figure (in his *Ars magna et ultima*) reveals how far behind he was of the geometry of his time.[12] His method does not provide even a close approximation of the perimeter or area of the desired square.[13]

Books on the Lullian art proliferated throughout the seventeenth century, many of them carrying inserted sheets of circles to be cut out, or actual rotating circles with centers attached permanently to the page. Wildly exaggerated claims were made for the method. The German Jesuit Athanasius Kircher (1601–1680), scientist, mathematician, cryptographer, and student of Egyptian hieroglyphics, was also a confirmed Lullist. He published in Amsterdam in 1669 a huge tome of nearly five hundred pages titled *Ars magna sciendi sive combinatoria*. It abounds with Lullian figures and circles bearing ingenious pictographic symbols of his own devising.[14]

The eighteenth century witnessed renewed opposition to Lull's teachings in Majorca and the publication of many Spanish books and pamphlets either attacking or defending him. Benito Feyjóo, in the second volume of his *Cartas eruditas y curiosas* ("Letters erudite and curious"), ridiculed Lull's art so effectively that he provoked a two-volume reply in 1749–1750 by the Cistercian monk Antonio Raymundo Pasqual, a professor of philosophy at the Lullian University of Majorca. This was followed in 1778 by Pasqual's *Vinciciae Lullianae*, an important early biography and defense of Lull. The nineteenth and twentieth centuries saw a gradual decline of interest in the Art and a corresponding increase of attention toward Lull as a poet and mystic. A periodical devoted to Lullian studies, the *Revista luliana*, flourished from 1901 to 1905. Today there are many enthusiastic admirers of Lull in Majorca and other parts of Spain, though the practice of his Art has all but completely vanished.

The Church has approved Lull's beatification, but there seems little likelihood he will ever be canonized. There are three principal reasons.

His books contain much that may be considered heretical. His martyrdom seems to have been provoked by such rash behavior that it takes on the coloration of a suicide. And finally, his insistence on the divine origin of his Art and his constant emphasis on its indispensability as a tool for the conversion of infidels lends a touch of madness, certainly of the fantastic, to Lull's personality.

Lull himself was fully aware that his life was a fantastic one. He even wrote a book called *The Dispute of a Cleric and Ramon the Fantastic* in which he and a priest each try to prove that the other has had the most preposterous life. At other times he speaks of himself as "Ramon the Fool." He was indeed a Spanish *joglar* of the faith, a troubadour who sang his passionate love songs to his Beloved and twirled his colored circles as a juggler twirls his colored plates, more to the amusement or annoyance of his countrymen than to their edification. No one need regret that the controversy over his Great Art has at last been laid to rest and that the world is free to admire Lull as the first great writer in the Catalan tongue, and a religious eccentric unique in medieval Spanish history.

NOTES

1. In later years Leibniz was often critical of Lull, but he always regarded as sound the basic project sketched in his *Dissertio de arte combinatorial*. In a letter written in 1714 he makes the following comments:

> When I was young, I found pleasure in the Lullian art, yet I thought also that I found some defects in it, and I said something about these in a little schoolboyish essay called *On the Art of Combinations*, published in 1666, and later reprinted without my permission. But I do not readily disdain anything—except the arts of divination, which are nothing but pure cheating—and I have found something valuable, too, in the art of Lully and in the *Digestum sapientiae* of the Capuchin, Father Ives, which pleased me greatly because he found a way to apply Lully's generalities to useful particular problems. But it seems to me that Descartes had a profundity of an entirely different level. (*Gottfried Wilhelm von Leibniz: Philosophical Papers and Letters*, edited and translated by Leroy E. Loemker, University of Chicago Press, 1956, vol. 2, p. 1067)

2. In sketching Lull's life I have relied almost entirely on E. Allison Peer's magnificent biography, *Ramon Lull*, London, 1929, the only adequate study of Lull in English. An earlier and briefer biography, *Raymond Lull, the Illuminated Doctor*, was published in London, 1904, by W. T. A. Barber, who also contributed an informative article on Lull to the *Encyclopedia of Religion and Ethics*. Other English references worth noting are: Otto Zöckler's article in the *Religious Encyclopedia*; William Turner's article in the *Catholic Encyclopedia*; George Sarton, *Introduction to the History of Science*, 1931, vol. II, pp. 900 ff.; and Lynn Thorndike, *A History of Magic and Experimental Science*, 1923, vol. II, pp. 862 ff.

A voluminous bibliography of Lull's works, with short summaries of each, may be found in the *Histoire littéraire de la France*, Paris, 1885, vol. XXIX, pp. 1–386, an indispensable reference for students of Lull. There also is an excellent article on Lull by P. Ephrem Langpré in vol. IX of the *Dictionnaire de théologie Catholique*, Paris, 1927. It is interesting to note that a 420-page novel based on the life of Lull, *Le Docteur illuminé*, by Lucien Graux, appeared in Paris in 1927.

The most accessible Spanish references are the articles on Lull in the *Encyclopedia universal ilustrada*, Barcelona, 1923, and vol. 1 of *Historia de la filosofia española*, by Tomás Carreras y Artau, Madrid, 1939.

3. Quoted by Peers, *Ramon Lull*, p. 64.

4. An English translation by Peers was published in 1926.

5. Separately issued in English translation by Peers in 1923.

6. Lull's death is the basis of a short story by Aldous Huxley, "The Death of Lully," in his book *Limbo*, 1921.

7. The only satisfactory description in English of Lull's method is in vol. 1 of Johann Erdmann's *History of Philosophy*, English translation, London, 1910. There are no English editions of any of Lull's books dealing with his Art. Peers's biography may be consulted for a list of Latin and Spanish editions of Lull's writings.

8. See *La logique de Leibniz*, by Louis Couturat, Paris, 1901, chap. IV, and *Leibniz*, by Ruth Lydia Shaw, London, 1954, chap. VIII.

9. Chapter 7 of my *Mathematics, Magic, and Mystery*, 1956, contains a reproduction and analysis of Loyd's "Get off the earth" puzzle and several related paradoxes.

10. *Historia del Lulisme*, by Joan Avinyó, a history of Lullism to the eighteenth century, was published in Barcelona in 1925. My quick survey of Lullism draws largely on Peers's account.

11. With the exception of the table of permutations, all these diagrams are reproduced and discussed in Zetzner's one-volume Latin edition of several of Lull's works, first printed in Strasbourg, 1598.

12. Bryson of Heraclea, a pupil of Socrates, had recognized that, if you keep increasing the number of sides of the inscribed and circumscribed polygons, you get increasingly closer approximations of the circle. It was through applying this method of limits that Archimedes was able to conclude that pi was somewhere between 3.141 and 3.142.

13. It has been called to my attention that, if a diagonal line *AB* is drawn on Lull's figure as shown in figure 15, it gives an extremely good approximation to the side of a square with an area equal to the area of the circle.

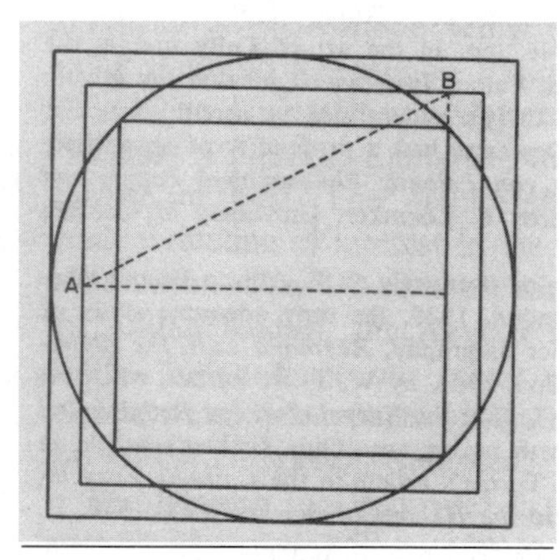

Figure 15.

14. Kircher's enormous books are fascinating mixtures of science and nonsense. He seems to have anticipated motion pictures by constructing a magic lantern that threw images on a screen in fairly rapid succession to illustrate such events as the ascension of Christ. He invented (as did Leibniz) an early calculating machine. On the other hand, he devoted a 250-page treatise to details in the construction of Noah's Ark!

Kircher's work on the Lullian art appeared three years after Leibniz's youthful treatise of similar title (see note 1). Leibniz later wrote that he had hoped to find important matters discussed in Kircher's book but was disappointed to discover that it "had merely revived the Lullian art or something similar to it, but that the author had not even dreamed of the true analysis of human thought." (Vol. 1, p. 352, of the edition of Leibniz's papers and letters cited in note 1.)

7.

THE BANACH-TARSKI PARADOX

No paradox in solid geometry is more mind-boggling than the Banach-Tarski paradox. Understanding it is a fine introduction to the strange properties of sets with an infinity of points. I know of no better explanation of this famous paradox, along with a detailed proof of its validity, than Leonard Wapner's *The Pea and the Sun* (A. K. Peters, 2005). The following review appeared in *New Criterion* (October 2005).

There is a big difference between what mathematicians call a fallacy and what they call a paradox. A fallacy is a flawed proof, such as the "proof" on page fifty-four of the book under review, that all triangles are isosceles. A paradox is as assertion almost impossible to believe but nevertheless true. A good example is the famous twin paradox of relativity. If one twin travels a long distance from the earth, at a fast speed, then returns, she'll be younger than her stay-at-home sister. The time difference can be arbitrarily large. If the traveling twin goes to a distant galaxy at a velocity near that of light, then returns, thousands of earth years could have gone by. Time travel into the far future (not into the past) is a genuine possibility!

The twin paradox, which incidentally has been empirically confirmed, is not hard to comprehend if one is familiar with the time dilation of relativity theory. Far more mind-blowing is a mathematical result known as the Banach-Tarski paradox after two Polish mathematicians, Stefan Banach and Alfred Tarski.

Tarski is best known for his semantic theory of truth. It eliminates such logical paradoxes as "This statement is false" by forbidding a language to talk about the truth or falsity of statements in the same language. Tarski's famous example is "Snow is white" is true if and only if snow is white. The sentence inside quotes is in a metalanguage asserting the truth of a statement in the language of physical realism. To talk about true or false in a metalanguage requires a meta-metalanguage, and so on into an infinite hierarchy of languages.

Tarski's way of defining truth has the great merit of applying both to the language of science and to languages of logic and mathematics. It played a key role in the fading of efforts by followers of Dewey and William James to define truth as the passing of tests for truth.

The original version of the Banach-Tarski (BT) paradox shows how a solid sphere can be cut into a finite number of point sets that can be shifted about to make two balls identical in shape and size to the original! Raphael Robinson, an American mathematician, reduced the required number of point sets to five—four, each with an infinity of points, and a single point from the original sphere's center.

Other forms of the BT paradox are even more counterintuitive. Each of the two magically created balls can be dissected into point sets that can be rearranged to make four balls, and this magnification can be continued to produce as many balls as one pleases. In a still crazier version, a tiny sphere can be cut into point sets that recombine to make a sphere of *any* size—hence the book's title, *The Pea and the Sun*. Even worse, the two objects may be of any size of shape. As the author puts it, an ideal mosquito can be transformed into an ideal elephant!

It must be said at once that such miraculous changes can only be done in theory. There is no way to slice an apple into parts that will make two apples or an elephant, no way, as Wapner writes, to multiply gold or enlarge loaves and fishes to feed starving millions. The BT paradoxes occur only within an abstract system of mathematics which deals with what are called "transfinite sets"—sets first recognized by the great German mathematician Georg Cantor.

Leonard Wapner, professor of mathematics at El Comino College in

Torrance, California, makes the BT paradox the centerpiece of a marvelous book (his first, by the way). Chapters proving the paradox are tough going for mathematically uninformed readers, but don't let that put you off if you are such a person. Wapner has surrounded his central theme with a wealth of easily understood topics, much of it recreational, related in some way to the BT paradox.

The book opens with a crystal-clear introduction to Cantor's infinite ladder of transfinite numbers. Cantor called the lowest number aleph-zero. It counts all rational numbers—the natural numbers 1, 2, 3, . . . and their integral fractions. Like all transfinite numbers, the set has the bizarre property that its points can be put into one-to-one correspondence with an infinite portion of itself. Even Galileo knew this curious property. He noticed, perhaps with surprise, that the counting numbers could be paired one to one with a subset such as the square numbers.

Of course, the counting numbers can be paired with even "smaller" infinite sets such as the prime numbers. Although in one sense there are far more counting numbers than squares, there is another sense in which their numbers are the same. This strange property underlies a story, well retold by Wapner, about tasks facing the manager of Hotel Infinity.

The hotel's endless numbers of rooms are all occupied when ten travelers arrive, each demanding a room. No problem. The manager simply shifts everyone from a room n to room $n + 10$. This leaves the first ten rooms vacant. Next week an infinity of travelers arrives, each wanting a room. Again, no problem. The clever manager moves each guest to a room with a number twice that of his former room. This opens all odd-numbered rooms.

Cantor was able to show that aleph-one, the next higher number to aleph-zero, counted sets that could *not* be put into one-to-one correspondence with the rational numbers. It was, therefore, a higher infinity. Cantor did this by an ingenious "diagonal" technique.

Aleph-one counts all the real numbers—the rationals plus the irrationals such as pi and the square root of 2. The number of points on a line is aleph-one. So is the number of points on a square, or a cube, or any solid object of higher dimensions. Aleph-two counts all the subsets of aleph-one. Aleph-three counts the subsets of aleph-two. Each aleph can be expressed by 2 raised to the power of its preceding aleph. The ladder of alephs is infinite, although beyond aleph-three the alephs have almost no mathematical uses.

A tantalizing question arises. Is it possible that there are transfinite numbers between two alephs, especially between aleph-zero and aleph-one? Cantor tried without success to prove there are no such in-betweens. As Wapner informs us, the question was not laid to rest until Stanford University's Paul Cohen proved that the problem is Gödel-undecidable within standard set theory. One may assume without contradiction that no such numbers exist, or one can assume an infinity of in-between numbers.

Cohen also startled his colleagues with another unexpected result. The BT paradox requires for its proof a notorious axiom called the "axiom of choice." It asserts that from any number of sets you can always select exactly one member of each set to create a new set. This obviously can be done with a finite or an infinite number of finite sets, but when each set is also infinite, severe difficulties arise.

Bertrand Russell, Wapner tells us, explained this by considering shoes and socks. From an infinite set of pairs of shoes you can take, say, a right shoe from each pair to form another infinite set. But if each set consists of an infinity of identical socks, there is no clear rule about how to select exactly one sock from each set. The axiom allows you to do this even though you can't specify exactly *how* to do it.

The BT paradox is one of many point-set paradoxes that cannot be proved without the axiom of choice. Cohen was able to show that a consistent set theory may include or exclude the axiom of choice. It remains a mysterious "existence" axiom independent of all the other axioms of standard set theory.

Although Cantor's alephs no longer worry today's mathematicians, Wapner cites several mathematicians of the past who regarded Cantor's alephs as mystical nonsense. Henri Poincaré called them a "malady, a perverse illness from which one day mathematics would be cured." Wapner also quotes Hermann Weyl's description of Cantor's ladder as "fog on fog." Leopold Kronecker, Cantor's former Berlin teacher, branded Cantor a "charlatan" and "corrupter of youth."

Cantor, Wapner reveals, was a devout Protestant who believed that his work was inspired by God. Outside of mathematics he held wild opinions, spending his later years trying to convince the world that Francis Bacon wrote all of Shakespeare's plays. After several nervous breakdowns, he ended his days in a mental facility.

Several chapters in Wapner's book cover entertaining topics that in some way resemble the BT paradox. One chapter is titled "Baby BTs." A variety of geometrical paradoxes are presented in which a polygon is cut into a small number of parts that can be reassembled to make another polygon with a different area! Paul Curry, an amateur Manhattan magician, was the first to show how pieces of a polygon could be rearranged to form a seemingly identical polygon with a hole!

Wapner reproduces puzzle maker Sam Loyd's bewildering paradox of a vanishing Chinese warrior. Fourteen warriors are around the rim of a disk. After a small rotation of the disk, one warrior vanishes. A linear version of the paradox, also reproduced by Wapner, involves fifteen leprechauns in a row. By switching two rectangular pieces, a leprechaun totally disappears. Which one vanished? And where did the little fellow go? In another chapter Wapner displays a number of beautiful dissections in which a polygon is cut into parts which reform to make a polygon of different shape. He outlines a lovely proof that any given polygon can be dissected into a finite number of pieces that will form any desired different polygon of the same area. There is a companion proof, described by Wapner, that similar transformations cannot be done with certain polyhedrons.

An old counterfeiting method is explained. It allows a thief to slice each of nine currency bills into two parts; then the parts can be rearranged to make ten bills. It's almost a BT paradox! Bill numbers on left and right sides, and top and bottom, are there precisely to foil this technique for multiplying currency.

After a detailed proof of the BT paradoxes, Wapner turns his attention to some speculations about the future of mathematics. He considers the implications of Moore's law, which predicts that every eighteen months computer power will double. The incredible speed of today's supercomputers has led to what Wapner calls "experimental mathematics." The computer has become a tool, like a telescope or a microscope, for testing and even suggesting conjectures. Occasionally a computer proof will require a printout so vast it can be checked only by another computer. Some computer programs will not validate a proof, but establish a result with only a very high degree of probability. The resemblance to science is obvious. Future computer "proofs" may end, Wapner jokingly writes, not with Q.E.D., but with "You can bet on it—trust me!"

Tomorrow's computers, Wapner believes, will be enormously faster.

They may use light instead of electricity to twiddle symbols. They may exploit the properties of DNA. Looming large on the horizon are quantum computers capable of speeds Wapner calls "astronomical."

Wapner ends his book with a profound question which science is nowhere close to answering: Will Cantor's alephs and the BT paradoxes ever find applications in the physical world? To my astonishment, Wapner introduces three scientists who are seriously speculating on just such applications. Two American physicists, Roger S. Jones and Bruno Augenstein, conjecture that the BT paradox may actually play a role in the behavior of hadrons! How can a muon be exactly like an electron except that it is larger, heavier, and short-lived? Has an electron been magnified by something akin to BT magnification?

The third scientist is the astrophysicist M. S. El Naschie, at the University of Cairo. Wapner lists two of his papers in his seven-page bibliography. Naschie wonders if a BT magnification may have produced the big bang. If the universe ever stops expanding, and goes the other way toward a big crunch, perhaps BT compression will be at work!

In considering such fantastic speculations it is good to realize that transfinite sets seem to "exist" only in the Platonic world of pure mathematics. Only in *that* world can a mosquito's infinity of points be put into one-to-one correspondence with the points of an elephant. What is so amazing is that in Plato's realm a mosquito can be cut into a *finite* number of parts that will reassemble to make an elephant. In the real world, of course, no material object has an infinity of points. Indeed, it has no points at all. Points exist only in formal mathematical systems. Matter is made of molecules, in turn made of atoms, in turn made of particles which could be vibrating loops of string. Almost all of a material object is empty space.

The "Go-Go Principle" permits anything to occur that is not logically forbidden. Wapner reminds us that both relativity and quantum mechanics bristle with paradoxes that are extreme violations of common sense but which are known to be true. Perhaps someday scientists may discover that Nature knows all about transfinite sets and BT paradoxes. On some level, far below quantum mechanics and possible superstrings, or in dimensions high above those we know, Einstein's Old One may be juggling Cantor's alephs in ways we cannot yet—perhaps never can—comprehend.

8.

TRANSCENDENTAL NUMBERS AND EARLY BIRDS

You can imagine my amazement when I first came upon Jamie Poniachik's discovery that pi to four decimals is hiding among the first dozen decimal digits of pi. This led to my suggestion that searching pi for what I called famous "early bird" numbers would be a pleasant, albeit useless, game. I introduced the pastime in the following article that appeared in *Math Horizons* (November 2005).

The concept of early birds intrigued the Russian mathematician Mamikon Mnatsakanian, now living in Pasadena. He has generalized early birds in several ways and made some nontrivial discoveries of theorems that he will be reporting in a forthcoming paper.

It's hard to believe that it was not until 1844 that transcendental numbers were known to exist! But first, a few definitions.

A *rational number* is one that can be written as *a/b*, where *a* and *b* are integers. In decimal form, rational numbers either terminate (1/4 = .25) or they have a pattern of consecutive digits that repeat endlessly (1/7 = .142857 142857 142857 . . .).

An *irrational number* is one that cannot be expressed as *a/b* where *a* and b are integers. In decimal form, it never ends, and it has no pattern of consecutive or otherwise digits that keep repeating.

An *algebraic number* is the root of a polynomial equation with rational coefficients. It can be rational or irrational. Thus 2 is algebraic because it is the root of such equations as $10x = 20$. The square root of 2 is algebraic because it is the root of $x^2 = 2$.

Transcendental numbers, such as π and e, are irrational numbers that are *not* the roots of algebraic equations with rational coefficients.

In 1844, the French mathematician Joseph Liouville (1809–1882) first proved the existence of transcendentals by actually constructing an infinite number of them. Liouville numbers, as they are known, can be called artificial because Liouville did not find them anywhere in mathematics. He simply made them up out of whole cloth.

The simplest Liouville number is the binary transcendental shown below:

$$.110001000000000000000001\ldots$$

The 1s are at positions given by consecutive factorials. Thus the first 1 is at position $1! = 1$, the second is at position $2! = 2$, the third is at position $3! = 6$, the fourth is at position $4! = 24$, and so on.

It's easy to construct artificial numbers that are obviously irrational but not transcendental (like the square root of 2). It is not so easy to construct such numbers that are transcendental. Consider

$$.101001000100001\ldots$$

Liouville showed that any number of this form, where the 1s can be replaced by any constant digit from 2 through 9, is transcendental.

It was not long until π, e, and scores of other famous irrationals were shown to be transcendental. In spite of great recent progress in proving numbers transcendental, many deep unsolved questions remain. For example, e^π has been proved transcendental, but no one yet knows if π^e is transcendental or even if $\pi + e$ or πe or e^e or π^π are irrational! All are believed to be transcendental, but until some genius proves otherwise, it is possible that the sum of π and e is a rational number!

Other than Liouville numbers, the most famous of all artificial transcendentals is made simply by putting a decimal point in front of the counting numbers:

$$.123456789101112131415\ldots$$

This is called Mahler's number, after Kurt Mahler, a mathematician who first proved that this counting number is transcendental in all bases. You will find a neat proof for base 10 in Ivan Niven's marvelous little book *Irrational Numbers*, recently republished by the Mathematical Association of America.

A number is *normal* if every digit, every doublet, every triplet, or any specified pattern of digits appear the number of times you would expect if the number were generated by a randomizer. So far, π, e, and the irrational roots of all algebraic numbers have passed all statistical tests for normalcy, but none has yet been proved normal.

In 1934, it was shown that a^b is transcendental if a is an algebraic number not 0 or 1, and b is any irrational number. Thus 2^π and $10^{\sqrt{2}}$ are transcendental.

Now for a startling coincidence. Note the last five digits of Mahler's number. They are π to four decimals! (I'm indebted to Jaime Poniachik, an Argentine puzzle maker, for telling me about this.)

Every positive integer obviously appears somewhere in Mahler's number, but many numbers show up far ahead of where they normally would occur as a counting number. Pi turns up so early because 3 is the last digit of 13, followed by 14 and 15. Poniachik points out that π to 5 decimals appears as early as 14159**3-14159**4. Consider 666. Of course it shows up in the counting number, but you can spot it early in the sequence 65-**66-6**7.

Much innocent but totally useless amusement can be had by searching for the earliest appearance of such familiar numbers as integral powers of roots, the last four digits of your phone number, your house number, and so on. Better still, see if you can formulate a procedure for determining the earliest appearance of any given integer in the counting numbers.

If a number appears ahead of its position as a counting number, let's call it an *early bird*. The famous year 1492, for another example, is an early bird (49**1-492**). Poniachek points out that 8192, the thirteenth power of 2, is an early bird (1**8-19-2**0). Like π, the transcendental number *e*, to four decimals (2.7182) is also an early bird (**2718-2**719), though not as early as π.

What is the smallest early bird prime? When will the next coming year be an early bird? Mathematician Solomon W. Golomb, in a personal letter, mentioned that his home zip code 91011 (9-10-11) is a very early bird. The early bird numbers, Golomb wrote, are a subset of the counting numbers, and therefore countably infinite. He is convinced that "almost all" integers are early birds. There are no one-digit early birds, but half (45 out of 90) two-digit numbers are early birds, with rapidly increasing members of the species as the number of their digits increases.

9.

A DEFENSE OF PLATONIC REALISM

The oldest conflict among philosophically inclined mathematicians is whether mathematical objects and theorems have a reality of some sort independent of human cultures, or whether their locus is entirely inside human skulls. Plato was the first great Platonist, and the term Platonism is still used today for what is more often called mathematical realism. Almost all great mathematicians of the past and present have been and are realists, but there are a few notable exceptions. They include E. Brian Davies, whose book *Science in the Looking Glass: What Do Scientists Really Know?* (Oxford, 2003) is here reviewed. The review first appeared in *Notices of the AMS* (vol. 52, September 2005).

This review gave me a chance to sound off once again in defense of mathematical realism. I continue to be amazed that any professional mathematician would suppose that mathematics has no reality apart from human cultures. I am even more astounded that there actually are physicists who think the moon would not be "out there" if no one (not even a mouse? Einstein liked to ask) observed it.

I'm not sure exactly what Brian Davies, a distinguished mathematician at King's College, London, intended the title of his fifth book to suggest. Reflect like a mirror the history and nature of science? Perhaps he also thought of his book as leading readers into a dreamlike universe as fantastic as the world Alice first entered through a rabbit hole and later through a looking glass. Whatever the intent, it is a brilliant work, beautifully written, and brimming with surprising information and stimulating philosophical speculations.

Before turning to my one caveat—unlike Davies I'm an unabashed realist who believes that mathematical objects and theorems are "out there" with a peculiar kind of reality that is independent of minds and cultures—let me go over some of the book's highlights.

Davies begins with a discussion of the uncertainties of perception. Errors of seeing are demonstrated with two amazing optical illusions. One is a ring of slash strokes that seems to rotate as the page is shifted forward and back. It is impossible, viewing the other illusion, not to be sure that one square of a checkerboard is darker than another when both are actually the same shade. Language too can be misleading. Davies does not buy Noam Chomsky's claim that there is a genetically transmitted deep "universal grammar." Interaction with environment is sufficient to account for the ability to speak. Apes failed to speak because their throats lacked the apparatus necessary for producing a great variety of sounds.

Descartes's famous effort to separate mind from body is thoroughly discredited. On the other hand, although Davies is convinced that consciousness—the awareness that one exists and has free will—is a function of a material brain, it is still a total mystery. He agrees with Roger Penrose, Oxford's mathematical physicist, that no computer made with wires and switches will ever become aware of what it is doing. Assuming that we are no more than an enormously complex pattern of molecules, Davies speculates on the possibility that our pattern could someday be scanned and transmitted to another place like the up and down beaming of characters in *Star Trek*. Can a simpler pattern, such as an apple, be so translocated?

Physicists are hard at work trying to accomplish just such a feat and have actually succeeded in teleporting an atom. Transmitting a human, however, as Davies recognizes, raises profound questions about identity. After being "beamed down," would a translocated person be the same person or merely a replica? What if the technique produces two identical persons? Philosophers, notably Locke, have agonized over just such thought experiments. Hundreds of science fiction tales have considered such possibilities. Penrose, by the way, has argued that if even an apple is transmitted, laws of quantum mechanics require total destruction of the original. When the captain of the *Enterprise* is beamed down to a planet, he cannot leave himself behind.

Davies's chapters on pure mathematics cover a wide range. He deals with imaginary and complex numbers and the difficulties that arise with rational numbers when they are enormously large or small. "Hard" problems like the four-color-map theorem have finally been proved, but by such monstrous computer printouts that the proof can be checked only by another computer. Goldbach's still unsettled conjecture that every even number greater than four is the sum of two odd primes has now been confirmed for numbers up to 10^{14}. This, Davies adds, "would be sufficient evidence for anyone except a mathematician."

A page is devoted to the notorious Collantz conjecture. Start with any number above 1. If even, halve it. If odd, replace it with $3n + 1$. Continue doing this. If the procedure ends with 1, stop. The conjecture is that it will always stop. So far it has stopped for all n up to 10^{12}, but a proof remains elusive.

Davies reports the sensational discovery a few years ago by Manindra Agrawal and his two young assistants in Kampur, India, of a simple rapid method of testing whether a huge number is or isn't prime. The algorithm doesn't generate factors, but merely tests for primality, and does so in polynomial time! "Such discoveries," Davies writes, "are among the things which make it a joy to be a mathematician."

Several pages concern the innocent-seeming little problem of the three doors, which created such a stir when Marilyn vos Savant published it in her weekly *Parade* column. Modeled with playing cards it goes like this. Smith places three cards face down on a table. Only one card is an ace. You are asked to guess where the ace is by placing a finger on a card. Clearly the probability you guess right is 1/3. Smith, who knows where the ace is, now turns face up a card that is *not* an ace. Two cards remain face down. Does not the probability your finger is on the ace go up to 1/2? It does not! It remains 1/3. If you now move your finger to the other card, the probability it rests on the ace rises to 2/3! Savant gave a correct solution, but thousands of mathematicians who should have known better wrote angry letters attacking the solution. The event even made the front page of the *New York Times*.

Davies's chapters on the physical and biological sciences are as broad in scope and as illuminating as his chapters on mathematics. We learn about the mind-bending paradoxes of relativity and quantum mechanics, about chaos theory, continental drift, and the ever-changing conjectures of cosmology, the anthropic principle, Thomas Kuhn's shaky views about science revolutions and paradigm shifts, and a hundred other topics on the frontiers of modern science.

A lengthy chapter on evolution rips apart the currently fashionable claim by defenders of "intelligent design" that the "irreducible complexity" of even the simplest life-forms could not have evolved without the guidance of an intelligent designer, namely, God. Davies ticks off a variety of facts that support the randomness of mutations. Why should a competent designer, he asks, bother to produce millions of dinosaurs only to allow them to vanish except for some small ones that turned into birds?

The world's vast amount of evil and suffering is evidence, Davies is convinced, that there is no transcendent deity supervising evolution. As Marlene Dietrich once remarked (my wording), "If there is a God, he must be crazy." Davies reproduces a lovely photograph of a snow crystal as evidence that natural laws combined with chance can produce intricate complexity.

I have touched on only a small fraction of the myriad of colorful accounts that Davies provides about today's science and mathematics. Let me now turn to my reasons for not accepting a basic theme of Davies's book. I refer to his constant bashing of mathematical realism, especially the vigorous Platonism of Penrose and Kurt Gödel.

First of all, I prefer the term realism to Platonism. Why? Because it avoids all the dismal controversies over such universals as goodness, beauty, chairness, cowness, and so on, that so agitated the minds of the medieval scholastics. No modern realist believes for a moment that numbers and theorems "exist" in the same way that stones and stars exist. Of course mathematical concepts are mental constructs and products of human culture. Everything persons think and do is part of culture. To say that numbers are mental constructs is to say something trivial—something no realist denies. The deeper question is whether these constructs have a peculiar, dimly understood kind of reality embedded in the universe in a way that is not mind-dependent. No human is needed to establish the fact that the geometrical shape of Aristotle's vase is inseparable from the vase. A spiral is inseparable from a spiral galaxy. The four corners of a cube can no more be detached from a physical model of a cube than from an ideal cube. The existence of optical illusions doesn't prevent one from seeing eight corners. You can close your eyes and feel the corners.

To a realist it is a misuse of language to say that primitive humans invented integers. What they did was invent *names*, later symbols, for properties of sets of discrete things such as fingers, pebbles, and elephants—things "out there," independent of human minds. Later they discovered the laws of arithmetic because that was how pebbles behaved when manipulated. They didn't invent the Pythagorean theorem. They found it, out there, when they measured the sides of material right triangles.

If one is a theist, believing as Paul Dirac did that God is a great mathematician, or even in the pantheistic deity of Spinoza and Einstein, then the locus of mathematical reality moves to a transcendent realm outside Plato's cave. The big debate between realism and constructivism evaporates. Paul Erdös liked to refer to God's *Book* in which all the most elegant proofs are recorded. From time to time mathematicians are permitted brief glimpses into one of the book's infinity of pages.

In a curious way, numbers may be *more* real than pebbles. Matter first dissolved into molecules, then into atoms, then into particles, which are now dissolving into tiny vibrating strings or maybe into Penrose's twistors. And what are strings and twistors made of? They are not made of anything except numbers. If so, the numbers are as much "out there" as molecules. They could be the *only* things out there. As a friend once said, the universe seems to be made of nothing, yet somehow it manages to exist. As Ron Graham remarked, mathematical structure may be the fundamental reality.

No antirealist such as Davies, and Reuben Hersh whom he admires, thinks the moon vanishes when no one, not even a mouse, is observing it. If the moon is "out there," why not admit that the moon's circumference, divided by its diameter, is a close approximation of $\neq$ even before mathematicians were around to say it and will be true if humans became as extinct as dinosaurs?

To an antirealist, π doesn't really exist outside the minds of sentient creatures. A sequence in π's decimal expansion, such as ten sevens in a row, isn't "there" until a computer calculates it. Davies tells an amusing story about how, in his book's first draft, he wondered whether a computer would ever find the sequence 0123456789 in π. To Davie's astonishment he later discovered that this sequence actually had been found. In the unlikely case that readers would like to know, the run starts at π's 17, 387,594,880th digit.

Davies takes up the question of whether one is allowed to say that somewhere in π is a run of a thousand sevens. In talking about such things, Davies, like all antirealists, slips into the language of realism. He writes, "we can estimate how long it would take to *find* the first occurrence" (italics mine) of a run of a thousand sevens. Again: The time it would probably take "to *find* the sequence" would be "vastly longer than the age of the universe." The word "find" of course implies that the run already exists. Davies is usually careful to avoid the word "find" because it gives the game away. "A Platonic mathematician would say that either there exists [such a run] . . . or there does not. This is certainly psychologically comfortable, but it is not necessary to accept it in order to be a mathematician." So comfortable, in fact, that antirealists seldom hesitate to speak of "finding" (i.e., discovering) something when they really mean constructing it.

William James somewhere speaks of digits as "sleeping" in π until some mathematician wakes them up. It is a striking metaphor. A sleeping cat, however, has to sleep somewhere. To Davies and Hersh the uncalculated digits of π sleep nowhere. They just pop into reality when a computer "constructs" them.

Bertrand Russell, a firm realist, once wrote that $2 + 2 = 4$ even in the interior of the sun. As I have often said, if two dinosaurs met two other dinosaurs in a clearing there would have been four there even if no humans were around to observe them. The equation $2 + 2 = 4$ is a timeless truth, valid in all logically possible worlds because it is what philosophers since Kant have called *analytic*. Given the axioms of arithmetic $2 + 2 = 4$ can be translated into a string of symbols which, assuming the axiomatic system's formation and transformation rules, arrive at $A = A$. Two plus two is four for the same reason that there are three feet in a yard.

Like many antirealists, Davies drifts close to a kind of social solipsism in which even the external world fades into a hazy construction of our brains. He quotes favorably from Donald Hoffman's book *Visual Intelligence: How We Create What We See*. "Why," Hoffman asks, "do we all see the same things?" Why, for instance, do we all see the same moon? Everyone I know would at once answer, "Because the moon doesn't change." Not Hoffman. His reason, so help me, is "because we all have the same rules of construction." We are not seeing a moon, out there, independent of us. We are seeing our constructions of the moon!

This is far more extreme than the opinion that $2 + 2 = 4$ because we all construct numbers the same way. To suppose that people see the same cow because they have constructed the cow by the same rules boggles my mind. They see the same cow because it *is* the same cow. "Realism," I once heard Russell say in a lecture, "is not a dirty word."

Antirealists are fond of claiming that mathematics, like science, is never certain. Morris Kline even wrote a book titled *Mathematics: The Loss of Certainty*. On the contrary, mathematics (including formal logic) is the *only* place where there is no loss of certainty. In his book *What Is Mathematics, Really?* Hersh argues that even laws of arithmetic are uncertain by considering a hotel that is missing a thirteenth floor. Take an elevator up eight floors, then go five floors more, and you reach floor fourteen. Hersh apparently thinks this violates the equation $8 + 5 = 13$. What he has done, of course, is jump from pure arithmetic to applied arithmetic, where applications are often uncertain.

Two beans plus two beans make four beans only if you assign to beans what Rudolf Carnap called a correspondence rule. In this case the rule is that each bean corresponds to 1. In the case of Hersh's elevator, if you assume that every floor corresponds to 1, then 8 floors plus 5 floors is sure to make 13 floors. Without correspondence rules, applications of mathematical truths are indeed uncertain. Two drops of water added to two drops can make a single drop. Hersh and Philip J. Davis, in their book *The Mathematical Experience*, give an even funnier example. A cup of milk, they inform us, added to a cup of popcorn doesn't make two cups of the mixture.

Euclidean geometry is not rendered uncertain because space-time is non-Euclidean. The Pythagorean theorem is absolutely certain within the formal system of plane geometry. There is not the slightest doubt that the angles of a Euclidean triangle add to 180 degrees. Science, on the other hand, is corrigible. Decades before Karl Popper, Charles Peirce coined the term fallibilism, and awareness that science is fallible goes back to the ancient Greek skeptics. As Hume taught us, there is no *logical* reason why the sun must rise tomorrow. For all we know there might be an unknown law of inertia that would suddenly stop the earth from rotating. This is in stark contrast with mathematics, where the uncertainty of science is incapable of inflicting injuries.

But enough about the tiresome, never-ending debate between the small minority of antirealists and the vast majority of mathematicians, including the greatest, who take realism for granted. They do their work without the slightest anxiety over the philosophical foundations of their craft.

What I admire most about Davies is his awe before the terrible mystery of time and why the universe, as Hawking recently wondered, "bothers to exist." He is aware of how little we understand the workings of Einstein's Old One. To answer his book's subtitle, scientists "really know" a great deal but what they don't know is even vaster. Here is how this book ends:

> The full complexity of reality is far beyond our ability to grasp, but our limited understanding has given us powers which we had no right to expect. There is no reason to believe that we are near the end of this road, and we may well hardly be past the beginning. The journey is what makes the enterprise fascinating. The fact that the full richness of the universe is beyond our limited comprehension makes it no less so.

The conflict between realists and their critics may come down finally to the choice of a language that is the least confusing. As President Clinton famously said, it all depends on what the meaning of *is* is.

ADDENDUM

In the same issue of the *Notices* in which my review appeared there is a paper by Davies titled "Whither Mathematics?" I sent to the AMS the following comments, but they were not printed.

I've just finished reading Brian Davies's stimulating article "Whither Mathematics?" Although I'm a hard-nosed realist (i.e., a neo-Platonist), and he is not, I can't disagree with his claim that mathematics took a sharp turn when Appel and Haken proved the four-color theorem with the help of a computer program so vast that it remained shaky for years. Since then there have been other "proofs" so lengthy and complex that there is still no consensus about their soundness. Hale's proof, for example, of Kepler's conjecture about the densest packing of unit spheres, or the classification of finite simple groups. As Davies tells us, the classification now fills five volumes with twelve yet to come!

Alas, much as I admire Davies's knowledge and writing skill, I must take issue with the conclusion he draws from these monstrous printouts. Like other antirealists he argues that the uncertainty of such efforts do serious damage to the ancient claim that mathematics is unique in possessing a certainty denied to science and history. It must now be seen as a fallible creation of "finite human beings," infused throughout with the same kind of doubts that infect science and all other human activities.

Now no one can deny that large areas of mathematics concern proofs that are suspect. This has always been the case, and always will be. It does not follow, however, as Davies insists, that "the difference between mathematics and other disciplines will be much reduced" and that "the unique status of mathematical entities will no longer seem relevant."

Except for unexplored jungles, almost all of mathematics consists of formal systems in which theorems are certain because they are *analytic*. They are tautologies generated by axioms, like the great truth, as Bertrand Russell liked to say, that there are three feet in a yard. Charles Peirce was fond of quoting his father Benjamin's definition of mathematics as the study of certain reasoning. The reasoning is certain precisely because it is independent of the physical world. As Einstein once said in a lecture, "as far as the propositions of mathematics refer to reality, they are not certain;

and as far as they are certain, they do not refer to reality." Except for a handful of extreme antirealists, no mathematician known to me doubts that 2 + 2 = 4 is a timeless truth, valid in all possible worlds. To quote Russell again, it is true even in the interior of the sun.

Nor is there the slightest doubt that the Pythagorean theorem follows inexorably from the symbols and rules of Euclidean geometry. I was surprised to find Davies flourishing the tiresome suggestion that general relativity somehow tarnishes Euclidean geometry. It does no such thing. That Einstein (a Platonist, by the way) found a non-Euclidean geometry applicable to space-time was an astonishing leap making possible a new theory of gravity, but it had not the slightest effect on any proof by Euclid.

Of course there are regions of mathematics, especially those involving infinity, where certainty has to be abandoned. In this respect mathematics does indeed resemble science. It also resembles science in other ways. Like physicists, mathematicians can experiment with numbers and models (such as diagrams on paper) and arrive at good conjectures. This experimental aspect of mathematics has been greatly augmented by computers. Not only do they assist in proofs, they also can give a conjecture, though possibly false, a probability of being true as high as one wants to go in a reasonable running time.

Davies is rightly impressed by a whopping note of uncertainty sounded by Gödel (an extreme Platonist) when he showed to the amazement of all mathematicians, especially Hilbert, that even as simple a formal system as arithmetic cannot prove itself consistent. Davies takes this to mean that in some far-off future a contradiction in arithmetic could be discovered! It might require, he writes, a proof of a hundred million pages, or one so complicated that we will never learn what Davies calls the "awful truth." Awful because if such a contradiction exists, it can be shown that *anything* could be proved!

I submit that this possibility—leaving aside the question of whether the consistency of arithmetic can be shown by going outside arithmetic—has no bearing whatever on endless beautiful ironclad theorems. Each is as certain as the knowledge that there are just five regular solids, that 1111111111111111111 is a prime, and that there are exactly 808,017,424,794,512,875,886,459,904,961,710,757,005,754,368,000,000,000 elements in the Monster group.

POSTSCRIPT ADDED IN 2006

One of the most extreme of recent attacks on mathematical realism is *Where Mathematics Comes From* (Basic Books, 2000), by George Lakoff and Rafael Nunez. They are cognitive scientists, not mathematicians. Mathematics, they maintain, is not "out there," in the universe; it is entirely the product of our brains. "All the 'fitting' between mathematics and the regularities of the physical world is done within the minds of physicists. . . . not in the regularities of the physical universe."

I found the above quote in Tom Siegfrid's entertaining book *Strange Matters* (Penguin Group, 2002). A realist, Siegfrid is as appalled as I am by the antirealism of Lakoff and Nunez. On pages 292–93 he quotes from a Lakoff lecture: "The only mathematics that we know is the mathematics that our brain allows us to know. So any question of math's being inherent in physical reality is moot, since there is no way to know whether it is or not."

Siegfrid quotes the contrary view of realist Murray Gell-Mann:

> "It seems that this whole theory is lurking there in some mathematical space," Gell-Mann said during a talk at Caltech in 2000. "It is there to be found. . . . The search for it appears to be a process of discovery, not invention. You are not adding bells and whistles in an effort to fit some empirical facts. You are gradually finding out what that preexisting self-consistent structure is."

"I don't think Gell-Mann would like Lakoff and Nunez's book," Siegfrid adds with understatement.

10.

THE JINN FROM HYPERSPACE

For several years I had the pleasure of writing a puzzle column in *Isaac Asimov's Science Fiction Magazine* in which I tried to embed a puzzle within a short science fiction yarn. The following tale ran in the July 1981 issue.

At that time Fermat's last theorem was the greatest unsolved conjecture in number theory. It was later validated in 1993 by Andrew Wiles with a proof so horrendous that no one could call it elegant. Will a simple and beautiful proof someday be discovered? Is it recorded in what Paul Erdös liked to call "God's Book"—a book in which all the best proofs are given? Or are there very simple theorems, such as Fermat's last or Goldbach's conjecture, that *have* no beautiful proofs? No one knows the answer to this question.

John Collier Fletcher had always wanted to be an opera star. He was a big man, but unfortunately his singing voice was on the small side—difficult for audiences to hear without electronic amplification. At college he gave up his dream, got a doctorate in mathematics, and became a professor at New York University. His specialty was number theory. For many years he struggled without success to prove Fermat's last theorem.

Fermat's last theorem asserts that the equation $a^n + b^n = c^n$ has no solution in positive integers if n is greater than 2. The case of $n = 1$ is trivial. When $n = 2$ there is an infinite number of solutions, called Pythagorean triples, of which the simplest is $3^2 + 4^2 = 5^2$. Pierre Fermat had made a note in the margin of a book saying he had a marvelous proof of this theorem, but that the margin was too small for it.

One wintry evening, when Fletcher was tramping through snow and slush to his bachelor's apartment in the SoHo (*So*uth of *Ho*uston) area near NYU, he passed a small store that he could not recall having seen before. A sign above the dirty window said: RAY PALMER'S OLD BOTTLE SHOP.

A shelf behind the window held a dozen or so curious bottles. One caught Fletcher's eye. It looked as if—yes, it surely was!—a Klein bottle.

(A Klein bottle is a closed surface without edges, like the surface of a sphere. A sphere's surface has two sides, outside and inside. An ant crawling outside cannot get inside unless there is a hole. But a Klein surface is one-sided like a Meobius strip. Outside is continuous with inside. Without going through a hole an ant can walk to any spot on both "sides" of the surface.)

Fletcher had always wanted to own a Klein bottle to show to his students. The shop's heavy door creaked ominously as he opened it. Through some tattered curtains at the back emerged an old man about four feet high, with white hair and watery blue eyes.

"Is that a Klein bottle in your window?" asked Fletcher.

"Well, not exactly," said the gnome. "It's just a crude model [see figure 1]. You'll observe that the stem goes through a hole where it enters the bottle. A true Klein bottle has no hole. The surface nowhere self-intersects because the stem twists around through the fourth dimension."

Figure 16. Model of Klein Bottle.

"I know, I know," said Fletcher. "I teach topology at New York University."

The gnome seemed unimpressed. "I do have a few genuine Klein bottles in stock. But they're more expensive. And they can be troublesome."

"Troublesome?" said Fletcher. "Why?"

"Because they twist through four-space. You never know what sort of creature from hyperspace might crawl out when you unstopper the bottle. It could be a friendly angel or jinn, but it might be something evil like a demon or a dero. Don't laugh. I'll show you one of the things."

Fletcher checked his laugh—actually it sounded more like a thin cackle—while the gnome disappeared behind the curtain. He emerged a moment later with a pear-shaped bottle almost as large as himself. It seemed to be made of rather fragile pink glass. It was a Klein bottle all right, except that where the stem usually plunged through a hole there was a spherical region of intense whiteness that shimmered and glowed like ball lightning. The gnome pointed to it with a black-edged fingernail

"That's where the miserable thing bends through hyperspace," he said. "Naturally you can't see the twist. But if you drop anything into the bottle, it will fall into the fourth dimension and you'll never recover it."

Fletcher was so intrigued that he bought the bottle at once even though it cost much more than he had anticipated. In his apartment he put the bottle in the center of his living room, then knelt on the rug beside it and tried to figure out what caused that cloud of scintillating light.

He tried to feel the cloud, but his hand simply vanished into a region of intense cold. When he removed his hand, his fingers were so frozen that he had to warm them under a hot-water faucet.

The opening at the top of the bottle was plugged by a black rubber stopper almost six inches across. By working it from side to side, Fletcher finally succeeded in pulling it out.

A loud popping sound was accompanied by a rush of icy air, a billowing cloud of purple smoke, and a strange Istanbul smell that seemed to mix sewage odors with aromatic spices. An enormous jinn, dressed as if he had popped straight out of *The Arabian Nights*, materialized from the cloud and made a low bow.

"I am at your command," he said in a deep resonant voice that Fletcher envied. "You have the usual three wishes. What is your first desire?"

After Fletcher recovered his composure he said hesitantly: "I've always wanted to sing like Caruso."

"To hear," said the jinn, "is to obey."

Fletcher felt a sudden surge of energy pulse through his lungs and it seemed as if his chest had enlarged several inches. He sang a few notes. Magnifico! The tone was perfect, the vibrato exquisite.

"Bravissimo!" said the jinn. "And your second wish?"

Fletcher thought for only a few seconds. "I would like a proof of Fermat's last theorem."

"I beg your pardon?" said the jinn.

Fletcher quickly scribbled an equation on a sheet of paper. "It's the greatest unsolved problem in number theory. If I can prove that this has no solution in integers when n exceeds 2, I'll be more famous than Isaac Asimov."

The jinn studied the equation. "I have a poor head for figures. This will require consultation with a higher authority. Don't leave. I'll be back in a few Earth minutes."

Somehow the jinn managed to flow into the pink bottle. A moment later, out he popped in another purple burst of strange-smelling smoke, and handed Fletcher the paper he had taken with him. Under the equation in small but legible handwriting, was a short proof.

Fletcher read the proof with mounting embarrassment. In his excitement he had written the wrong equation! He had interchanged the ns with a, b, and c. The equation had, so to speak, been inverted like this:

$$n^a + n^b = n^c$$

The easy-to-follow proof of impossibility when n exceeded 2 was certainly watertight. Can you devise such a proof?

Answer

I am indebted to Douglas Hofstadter for his proof of the upside-down version of Fermat's last theorem. Hofstadter introduces the theorem in his Pulitzer Prize–winning book *Gödel, Escher, Bach: An Eternal Golden Braid*. No proof is there given because, as the text says, the marvelous proof "is so small that it would be well-nigh invisible if written in the margin."

The equation $n^a + n^b = n^c$ obviously has no solution in positive integers if $n = 1$ because it then reduces to the false equality $1 + 1 = 1$. It has an infinity of solutions if $n = 2$. We have only to let $a = b$, and $c = a + 1$. For example: $2^2 + 2^2 = 2^3$.

Suppose n is greater than 2. If n is the base of a number notation, then all powers of n have an n-ary representation that is 1 followed by a string of 0s. Thus in our base-10 notation, all powers of 10 have the form: 10, 100, 1000, 10000, and so on.

In the upside-down equation either $a = b$ or a does not equal b. If $a = b$, the sum of the two equal powers, written in base-n notation, will be the sum of two numbers, each written as 1 followed by a zeros. The sum will have the form of 2 followed by a zeros, which obviously cannot be a power of n.

Suppose a is not equal to b. Each power will be written in base-n notation as a 1 followed by a string of 0s, but now the strings will be of different lengths. Therefore the sum will have the form of 1 followed by a string of 0s that will contain another 1 somewhere in the string. Once more, a number of this form cannot be a power of n. Since a must either equal or not equal b, we have proved the theorem by *reductio ad absurdum*. (You may wonder why this proof doesn't apply to binary notation when $a = b$, but a little experimentation will make it clear.)

The jinn scowled while Fletcher explained his mistake. "We did supply what you requested," he said. "Therefore it must count as a fulfillment of your second wish. What is your third?"

"I desire a proof that *this* equation," Fletcher boomed in his new stentorian voice, "has no solution when n is greater than 2!" This time he wrote the equation correctly.

"To hear," said the jinn, bowing, "is to obey. But I must again check with my superiors."

The jinn flowed back into the bottle. Several minutes passed. Fletcher could not stifle another impulse to test his voice. He sang a familiar aria from *Rigoletto*, ending on a high note. He belted out the note with all the lung power he could muster.

The pink bottle shattered into a thousand pieces.

Of course Fletcher never saw the jinn again. Nor could he locate the old bottle shop. It seemed to have vanished as completely as the jinn.

Fletcher changed his name and occupation. Perhaps you have heard of John Luciano Pavoletti, said to be the greatest tenor since Caruso.

Admirers of the fantasy of John Collier may recall his classic short story "Bottle Party," about a jinn and an unfortunate fellow named Frank Fletcher. If you are interested in a simple way to construct a Klein bottle out of paper, see chapter 2 of my *Sixth Book of Mathematical Games from Scientific American*. You can cut the paper model in half to discover that the two halves are Moebius strips, each a mirror image of the other.

11.

SATAN AND THE APPLE

I've been fascinated all my life by logical paradoxes. It is one reason why I am so fond of Lewis Carroll's fiction, especially his two Alice books and his *Sylvie and Bruno* novel. They abound in logical contradictions. The selection that follows is another of the puzzles I contributed to *Isaac Asimov's Science Fiction Magazine*. It appeared in the January 1985 issue.

I had occasion recently to spend a few weeks in San Francisco. Early one afternoon, while I was finishing lunch at the Caffe Puccini in the North Beach section—it's a spot where some of my more eccentric Silicon Valley friends hang out—an attractive young woman with red hair came over to the booth where I was sitting alone.

"Are you Martin Gardner?" she asked.

"That's me," I replied, putting aside the newspaper I had been reading.

"May I sit down?"

"Of course."

A mutual friend, she said, had told her I would be there. She was a subscriber to *Isaac Asimov's Science Fiction Magazine*, and she read my puzzle column every month. There was a strange story she wanted to tell me. She was sure I could make use of it in my column.

"It all started three years ago," she said, "a few weeks after I got my PhD in astrophysics at Stanford. My thesis was on the latest models of the inflationary universe—you know, the universe that starts with a big bang, then instantly inflates, and . . ."

"You don't have to explain," I interrupted. "I know about the models."

"Well," she continued, "I was sitting at the console of my Apple computer one night—it was late, about three in the morning—when I started to doze. I let my fingers wander idly over the keys . . ."

"Were you weary and ill at ease?" I asked. "Like the lady at the organ when she found that lost chord?"

"Exactly," she said. "That's how it was. I can't remember what I was thinking or dreaming about, but suddenly there was a loud explosion. The room filled with smoke. It smelled terrible. When the smoke cleared, there was Satan leering at me."

I couldn't help grinning. "What did the old fellow look like?"

"Just like his pictures," she said. "He was tall and handsome, wearing red tights, and he had a black mustache and pointed beard. There were two little horns on his head, and a long forked tail at his rear. He told me my fingers had hit on an old cabalistic combination of numbers and letters that forced him to appear. He offered me anything I desired."

"Of course you refused."

"No," she said. "I was too anxious to learn how the cosmos began. My thesis analyzed all the latest big bang models, but naturally I didn't know which was true. Maybe they were all false. I wanted to know what really happened back there in time some fifteen or twenty billion years ago."

"Did he tell you?"

"He did. And I must say his answer floored me. Maybe I shouldn't have been so surprised. After all, if it wasn't for the Bible we wouldn't know Satan existed. Right?"

"Right," I said. "What the devil did he tell you?"

"He told me the universe was formed just like it says in Genesis. Jehovah created everything out of nothing in six days, each twenty-four hours long. Then he rested on the seventh."

"If that's the case," said I, "the universe must be only six to ten thousand years old. What about the stars so far away that the light we see from them must have started millions of years ago?"

"I asked Satan about that," she said. "He told me the universe was created with all that light already on the way."

"And the fossils?"

"They're the records of plants and animals that were wiped out by Noah's flood."

"Are you trying to tell me," I said, "that the fundamentalists are right? That evolution is a false theory, and everything happened in six literal days just the way the Old Testament has it?"

"Yes," she said sadly. "That's what I learned from Satan. And he ought to know. He was there when it happened."

I studied her face carefully. It was a pretty face, with intelligent, unshifty eyes that looked directly into mine and betrayed not the slightest trace of insincerity. A sudden though popped into my head.

"Satan," I said, "has a reputation of never giving anything away free. There's always a bargain. You must have offered him something. What did you give in exchange for this information?"

She smiled faintly, looked nervously around the room, then bent over and whispered. "I gave up the ability of ever telling the truth when I meet someone for the first time."

The story (the plot of which, by the way, I snitched from Lord Dunsany's Story "Told Under Oath" in *The Ghosts of the Heaviside Layer*) is not strictly contradictory, but it has a strong flavor of self-referential paradox. If the woman always tells the truth, then her last sentence must be true. But if it's true, her entire story, including the final sentence, must be false. On the other hand, if the last sentence is a lie, perhaps her story is true.

Many novels, short stories, and poems have played with similar themes of self-reference. You'll find some of the classics discussed in the chapter on logic paradoxes in my book *Order and Surprise*. My favorite example is a limerick, but to understand it you must first contemplate the following two-liner:

There was a young man from Peru
Whose limericks stopped on line two.

What do you make of this shorter limerick?

There was a young man from Verdun.

Recalling the two-line limerick, your mind completes the second one with:

> Whose limericks stopped on line one.

Unfortunately, if you finish the verse this way, you give it *two* lines, thereby injecting a whopping contradiction.

Now for a final paradox. There is a certain event that I guarantee will or will not take place during the next ten minutes. You are absolutely incapable of predicting correctly whether it will or won't occur. I don't mean that it's *unlikely* you can predict it. I mean it is *logically impossible* to predict it!

You don't believe it? Then do the following. If you think the event will occur, write "Yes" inside the blank rectangle below. If you think it won't happen, write "No" inside the rectangle.

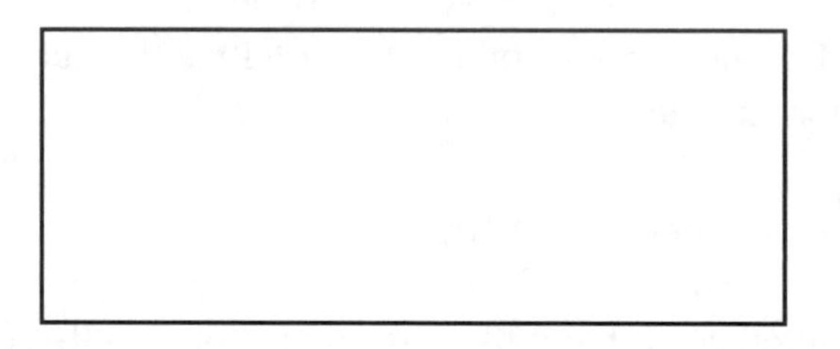

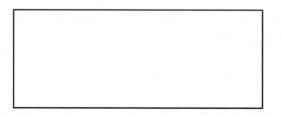

If you predicted correctly, I'll send you a million dollars.

The event is: You will write "No" inside the rectangle.

I introduced this version of a well-known prediction paradox in chapter 11 of *New Mathematical Diversions from Scientific American* (1966). It is one of the simplest of many prediction paradoxes that can arise whenever a prediction is causally related to the event being predicted. It can be further simplified by asking someone to reply yes or no to your question: "Will your reply be no?"

For a discussion of two famous prediction paradoxes, much harder to analyze than the one given here, see "The Paradox of the Unexpected Hanging," in my *Unexpected Hanging and Other Mathematical Diversions* (Simon & Schuster, 1969), and the chapter that follows this one.

12.

BLABBAGE'S DECISION PARADOX

Blabbage is a poor pun on Charles Babbage, England's pioneer computer scientist. There still is no agreement on how to resolve William Newcomb's powerful paradox, the topic of this chapter. I side with those who think the outright contradiction between the two decision procedures proves that the prediction machine, or a superbeing with similar powers, cannot exist.

Decision paradoxes do not arise when predictions are kept from the decider. For example, there is no paradox if a god, who knows the future, predicts that I will take a shower tonight. But if I am aware of the prediction, I can of course decide *not* to take a shower. Only when predictions interact in some way with a decider can logical paradoxes arise.

I first encountered Newcomb's paradox when I read Robert Nozick's paper about it in a book of essays honoring philosopher Carl Hempel. My *Scientific American* column on the paradox generated more than five hundred letters, almost equally split in defending one of the two decisions. The column, followed by Nozick's guest column commenting on the letters, can be found in *Knotted Doughnuts and Other Mathematical Entertainments* (W. H. Freeman, 1986). The following article was my November 1980 puzzle column in *Isaac Asimov's Science Fiction Magazine*.

After twenty years of work, Professor Charles Blabbage finally perfected his notorious decision prediction machine. Working details are too technical to explain, but essentially the device scans a human brain with three mutually perpendicular neutrino beams. Information on all electrical activity inside the skull is then analyzed by a powerful bubble computer. When any person is faced with a decision between two mutually exclusive courses of action, the machine can predict with amazing accuracy how he or she will decide.

For several months Professor Blabbage had been working with an amiable subject named Robert Zonick, obtaining an average success of 98 percent for all predictions.

"I have a new and curious test for you today, Bob," said the professor. "Observe that there are two boxes here on the table—one transparent, one opaque."

Bob nodded as he took his usual seat beside the table. Blabbage moved the three neutrino guns to within a few centimeters of Bob's forehead, his left temple, and the crown of his head.

"As you can see," Blabbage continued, "there is a hundred-dollar bill inside the transparent box."

"And the opaque box?" Bob asked, pointing his finger.

"At the moment it's empty," said the professor. "But let me explain." He glanced at his wristwatch. "One hour from now I'll ask you to make one of two choices. Either choose the opaque box only, or choose both boxes. If my machine predicts that you will take the opaque box, I'll put inside it a cashier's check for a million dollars. The money will be yours."

"Marvelous!" grinned Bob. "This test I like!"

"However, if my machine predicts you will take both boxes, I'll put *nothing* in the opaque box. Of course, you'll be certain then to get the hundred dollars."

Professor Blabbage pushed a button and the machine buzzed for a few seconds. Then he picked up the opaque box and left the room. A half hour later he returned to put the box on the table beside the transparent one.

"The computer has determined how you'll probably decide," he said.

"But think it over carefully. You have twenty minutes to make up your mind. Of course you must not touch either box until you make your choice. Everything is being videotaped. If the opaque box is empty now, it will be empty then. If it has the check inside it now, it will be there when you open it. Good luck, my friend."

After Blabbage left, Bob stared at the boxes for several minutes. "I've been tested a hundred times with this infernal machine," he said to himself, "and it was almost always right. So I should take only the opaque box. The odds are better than 9 to 1 that I'll get the big check. On the other hand . . ."

Bob suddenly realized that there was just as good an argument, maybe one even better, for taking *both* boxes! What is the argument?

Bob said to himself: "There are just two possibilities. The opaque box is either empty or it contains the check. Suppose it's empty. If I take only the opaque box I get nothing. But if I take both boxes I get at least a hundred dollars. Suppose the opaque box is not empty. If I take only it, I get the million-dollar check. But if I take both boxes I get the check *plus* a hundred dollars. Either way I'm sure to come out a hundred dollars ahead by taking both boxes!"

Each argument seems impeccable. According to experts on decision theory, which one is right?

Experts disagree! Some favor the "pragmatic argument" (take only the opaque box). Some favor the "logical argument" (take both boxes). Some say the paradox is not yet resolved. Still others maintain that the paradox proves the impossibility of prediction machines that work with better than 50 percent accuracy.

The problem is known as "Newcomb's paradox" after the American physicist William A. Newcomb who invented it in 1960. Zonick is an anagram of Nozick. It was Robert Nozick, a philosopher at Harvard, who first wrote about the paradox, and who contributed a guest column on it to the Mathematical Games department of *Scientific American* (see the fifth entry in the postscript's list of references). Nozick's column surveys the hundreds of letters from readers, including one from Asimov, who sought to resolve the paradox after I discussed it in an earlier column.

POSTSCRIPT

If you care to read some of the growing literature on this bewildering paradox, here is a chronological list of selected references:

Nozick, Robert. "Newcomb's Problem and Two Principles of Choice." In *Essays in Honor of Carl G. Hempel*, edited by Nicholas Rescher, 1970.

Howard, Nigel. *Paradoxes of Rationality: Theory of Metagames and Political Behavior*, 1971, pp. 168–84.

Bar-Hillel, Maya, and Avishai Margalit. "Newcomb's Paradox Revisited." *British Journal for the Philosophy of Science* 23 (November 1972).

Gardner, Martin. "Free Will Revisited." Mathematical Games Department, *Scientific American* (July 1973): 104–109.

Nozick, Robert. "Reflections on Newcomb's Problem." Mathematical Games Department, *Scientific American* (March 1974): 102–107.

Schlesinger, G. "The Unpredictability of Free Choices." *British Journal for the Philosophy of Science* 25 (September 1974).

Levi, Isaac. "Newcomb's Many Problems." *Theory and Decision* 6 (May 1975).

Brams, Steven J. "Newcomb's Problem and Prisoners' Dilemma." *Journal of Conflict Resolution* 19 (December 1975).

Brams, Steven J. "A Paradox of Prediction." *Paradoxes in Politics* (ch. 8), 1976.

Locke, Don. "How to Make a Newcomb Choice." *Analysis* 38 (January 1978).

Lewis, Daviod. "Prisoners' Dilemma Is a Newcomb Problem." *Philosophy and Public Affairs* 8 (Spring 1979).

Note: The two *Scientific American* columns are reprinted in my *Knotted Doughtnuts and Other Mathematical Bewilderments* (W. H. Freeman, 1986).

13.

PROFESSOR CRACKER'S ANTITELEPHONE

Tachyons were taken seriously for a while (they were an unwelcome aspect of early superstring theory), but today's physicists are convinced they do not exist. My tale about their use in a telephone appeared in the February 1980 issue of *Isaac Asimov's Science Fiction Magazine*.

Ada Loveface plays on the name of Ada Lovelace, Charles Babbage's lovely assistant and a daughter of Lord Byron. Cracker's name alludes to Alexander Graham Bell and to a popular bread product called a Graham Cracker. Barkback puns on Birkbeck, a London college.

Alexander Graham Cracker, a famous astrophysicist at Barkback College in London, was hard at work on perfecting a machine that could send signals through interstellar space at speeds faster than light. Back in the twentieth century, physicist Gerald Feinberg and others had found that relativity theory permits the existence of particles that always go faster than light. Feinberg called them "tachyons" after a Greek word for "swift."

Just as ordinary particles ("tardyons") can never be accelerated to the speed of light, so tachyons can never be slowed down to the speed of light. Since tachyons are always moving, they have no rest mass. This allowed Feinberg to represent their rest masses by imaginary numbers. After six months of intensive research, Professor Cracker finally designed what he called a "tachyonic antitelephone." Although tachyons had not yet been proved to exist, *if* they existed Cracker's antitelephone could modulate a beam of tachyons in such a way that signals could be carried by the beam.

"Because tachyons move faster than light," said Dr. Ada Loveface, the professor's assistant, "doesn't that mean they travel backward in time?"

"Of course," replied Professor Cracker. "Einstein's equations guarantee it. That's what's so marvelous about my antitelephone. We can send a message to intelligent aliens in the Andromeda galaxy at such a speed that it gets there several days before we send it!"

"In that case," said Dr. Loveface, "your antitelephone won't work."

Dr. Loveface then outlined a logical proof of her statement that was so ironclad that Professor Cracker abandoned his project at once. What sort of proof did she give?

Here is how Ada Loveface proved that if tachyons exist they can't be used for sending signals with speeds faster than light:

Suppose that *A*, at Barkback College, is in communication by tachyonic antitelephone with *B*, who lives on a planet at the other side of our galaxy. The tachyon speeds and the distance are such that if *A* sends a signal to *B*, and *B* instantly replies, then *B*'s signal will arrive one hour before *A* sends his signal. *A* could then get an answer to a question an hour before he asked it!

Dr. Loveface sharpened the contradiction as follows. Suppose *A* and *B* agree that *A* will ask his question at noon if and only if *B*'s immediate reply does *not* reach him by 11 AM on the same day. We are forced to conclude that an exchange of messages will take place if and only if it does not take place—a flat logical contradiction.

A few days after Professor Cracker had abandoned his project, Ada approached him and said: "Perhaps I was too hasty the other day in saying your antitelephone couldn't work. I've been reading some old science fiction classics about time travel, and they have suggested a possible way a modulated tachyon beam could send a signal that wouldn't lead to an absurdity."

What did Ada have in mind?

Dr. Loveface had read some old SF stories about time travel into the past that got around the familiar difficulty of whether a person would or wouldn't exist if he entered his past and killed his parents when they were babies. The gimmick was to assume that whenever anything from the future enters the past in a way that changes the past, the universe splits into two parallel worlds that are identical except that in one the alteration took place, in the other it didn't.

This gimmick can be applied to tachyonic messages. As soon as such a message enters the universe's past, the big split occurs. A person sending such a message can never get a reply because he continues to exist in the universe in which the message was sent, not in the duplicate universe in which the message was received. This permits one-way communication without contradiction, but not an exchange of tachyonic messages within the same universe.

POSTSCRIPT

The tachyon telephone is closely related to time travel paradoxes. You'll find these paradoxes discussed, together with the telephone, in my *Scientific American* column for May 1974, reprinted in *Time Travel and Other Mathematical Bewilderments* (W. H. Freeman, 1988). On the telephone paradox see "The Tachyonic Antitelephone," by G. A. Benford, D. L. Book, and W. A. Newcomb, in *Physical Review D*2 (July 15, 1970): 263–65. See also the excellent entries on "Tachyons," "Time Travel," and "Time Paradoxes," in *The Science Fiction Encyclopedia*, edited by Peter Nicholls.

An excellent introduction to tachyon theory is Gerald Feinberg's article on "Particles that Go Faster than Light," *Scientific American* (February 1970). Some parapsychologists have suggested that tachyons (if they exist) might be carriers of precognitive ESP (if it exists).

14.

ENERGY FROM THE VACUUM

I'm indebted to famous magician and baloney buster James Randi for passing on to me a copy of Colonel Bearden's fantastic treatise after he (Randi) had written an entertaining review of the book for his popular Web site. My review appeared in the January/February 2007 issue of *Skeptical Inquirer*. For an earlier discussion of Harold Puthoff's parallel search for a way to tap the energy of space see chapter 7 of *Did Adam and Eve Have Navels?* (Norton, 2000).

One of the strangest books ever written about modern physics was published here in 2002, and reprinted two years later. Titled *Energy from the Vacuum*, this monstrosity is two inches thick and weighs three pounds. Its title page lists the author as "Lt. Col. Thomas E. Bearden, Ph.D. (U.S. Army retired)."

"Dr." Bearden is fond of putting PhD after his name. An Internet check revealed that his doctorate was given, in his own words, for "life experience and life accomplishment." It was purchased from a diploma mill called Trinity College and University—a British institution with no building, campus, faculty, or president, and run from a post office box in Sioux Falls, South Dakota. The institution's owner, one Albert Wainwright, calls himself the college's "registrant."

Bearden's central message is clear and simple. He is persuaded that it is possible to extract unlimited free energy from the vacuum of spacetime. Indeed, he believes the world is on the brink of its greatest technological revolution. Forget about nuclear reactors. Vacuum energy will rescue us from global warming, eliminate poverty, and provide boundless clean energy for humanity's glorious future. All that is needed now is for the scientific community to abandon its "ostrich position" and allow adequate funding to Bearden and his associates.

To almost all physicists this quest for what is called "zero-point energy" (ZPE) is as hopeless as past efforts to build perpetual motion machines. Such skepticism drives Bearden up a wall. Only monumental ignorance, he writes, could prompt such criticism.

The nation's number two drumbeater for ZPE is none other than Harold Puthoff, who runs a think tank in Austin, Texas, where efforts to tap ZPE have been under way for years. In December 1997, to its shame, *Scientific American* ran an article praising Puthoff for his efforts. Nowhere did this article mention his dreary past.

Puthoff began his career as a dedicated Scientologist. He had been declared a "clear"—that is, a person free of malicious "engrams" recorded on his brain while he was an embryo. At Stanford Research International Puthoff and his then friend Russell Targ claimed to have validated "remote viewing" (a new name for distant clairvoyance), and also the great psi powers of Uri Geller. (See my chapter on Puthoff's search for ZPE in *Did Adam and Eve Have Navels?*, Norton, 2000.)

Bearden sprinkles his massive volume with admirable quotations from top physicists, past and present, occasionally correcting mistakes made by Einstein and others. For example, Bearden believes that the graviton moves much faster than the speed of light. He praises the work of almost every counterculture physicist of recent decades. He admires David Bohm's "quantum potential" and Mendel Sach's unified field theory. Oliver Heaviside and Nikola Tesla are two of his heroes. . . .

Bearden devotes several chapters to antigravity machines. Here is a sample of his views:

> In our approach to antigravity, one way to approach the problem is to have the mechanical apparatus also the source of an intense *negative energy* EM field, producing an intense flux of Dirac sea holes into and

> in the local surrounding spacetime. The excess charge removed from the Dirac holes can in fact be used in the electrical powering of the physical system, as was demonstrated in the Sweet VTA antigravity test. Then movements of the mechanical parts could involve movement of strong negative energy fields, hence strong curves of local spacetime that are local *strong negative gravity fields*. Or, better yet, movement of the charges themselves will also produce field-induced movement of the Dirac sea hole negative energy. This appears to be a practical method to manipulate the metric itself, along the lines proposed by Puthoff et al.[217]

The 217 refers to a footnote about a 2002 paper by Puthoff and two friends on how to use the vacuum field to power spacecraft. Bearden's antigravity propulsion system is neatly diagrammed on page 319. "Negatively charged local spacetime," says the diagram, "acts back upon source vehicle producing anti-gravity and unilateral thrust."

In the 1950s numerous distinguished writers, artists, and even philosophers (e.g., Paul Goodman, William Steig, and Paul Edwards) sat nude in Wilhelm Reich's "orgone accumulators" to absorb the healing rays of "orgone energy" coming from outer space. Bearden suspects (in footnote 78) that orgone energy "is really the transduction of the time-polarized photon energy into normal photon energy. We are assured by quantum field theory and the great negentropy solution to the source charge problem that the instantaneous scalar potential involves this process." I doubt if the Reichians, who are still around, will find this illuminating.

To my amazement Bearden has good things to say about the notorious "Dean drive"—a rotary motion device designed to propel spaceships by inertia. It was promoted by John Campbell when he edited *Astounding Science Fiction*, a magazine that unleashed L. Ron Hubbard's dianetics on a gullible public and made Hubbard a millionaire. Only elementary physics is needed to show that no inertial drive can move a spaceship in frictionless space. On pages 448–53 Bearden lists eighty patents for inertial drives. They have one feature in common. None of them works.

Counterculture scientists tend to be bitter over the "establishment's" inability to recognize their genius. Was not Galileo, they like to repeat, persecuted for his great discoveries? This bitterness is sometimes accompanied by paranoid fears, not just of conspiracies to silence them, but also fears of being murdered. Bearden's pages 406–53 are devoted to just such delusions.

Several kinds of "shooters" are described that induce fatal heart attacks. He himself, Bearden writes, has been hit by such devices. An associate, Stan Meyer, died after a "possible" hit by a close-range shooter. Another ZPE researcher was killed by a bazooka-size shooter. Steve Marikov, still another researcher, was assaulted by a sophisticated shooter and his body thrown off a rooftop to make it appear a suicide. When his body was removed, the pavement "glowed."

One day at a Texas airport a person three feet from Bearden was killed with symptoms suggesting he was murdered by an ice-dart dipped in curare! "That was apparently just to teach me 'they' were serious." The colonel goes on to explain that "they" refers to a "High Cabal" who were offended by a friend's "successful transmutation of copper (and other things) into gold. . . . We have had numerous other assassination attempts, too numerous to reiterate. . . . Over the years probably as many as 50 or more overunity researchers and inventors have been assassinated . . . some have simply disappeared abruptly and never have been heard from since." "Overunity" is Bearden's term for machines with energy outputs that exceed energy inputs.

> Any significant researcher should be wary of "meeting with a sudden suicide" on the way to the supermarket. Another thing to beware of, is a calibrated auto accident where your car is rammed from the rear, and you are shaken up considerably. An ambulance just happens to be passing by moments later, and it will take you to the hospital. If still conscious, the researcher must not get in the ambulance unless accompanied by a watchful friend who understands the situation and the danger. Otherwise, he can easily get a syringe of air into his veins, which will effectively turn him into a human vegetable. If he goes to the hospital safely, he must be guarded by friends day and night, for the same reason, else he runs a high risk of the "air syringe" assassination during the night.
>
> Simply trying to do scientific work, I find it necessary to often carry (legally) a hidden weapon. Both my wife and I have gun permits, and we frequently and legally carry concealed weapons.

> As early as the 1930s, T. Henry Moray—who built a successful COP>1.0 power system outputting 50 kW from a 55 lb power unit—had to ride in a bulletproof car in Salt Lake City, Utah. He was repeatedly fired at by snipers from the buildings or sidewalk, with the bullets sometimes sticking in the glass. He was also shot by a would-be assassin in his own laboratory, but overpowered his assassin and recovered.

Obviously I'm not competent to wade through Bearden's almost a thousand pages to point out what physicists tell me are howlers. I leave that task to experts, though I suspect very few will consider it worth their time even to read the book. To me, a mere science journalist, the book's dense, pompous jargon sounds like hilarious technical double-talk. The book's annotated glossary runs to more than 120 pages. There are 305 footnotes, 754 endnotes, and a valuable 73-page index.

A back cover calls the book "the definitive energy book of the 21st century." In my opinion it is destined to be the greatest work of outlandish science in both this and the previous century. It is much funnier, for instance, than Frank Tipler's best seller of a few decades ago, *The Physics of Immortality*, though not as funny as its 2007 sequel, *The Physics of Christianity*.

15.

PopCo

Serious mathematics has a central role in several recent novels and plays, and is currently involved in a weekly television series called *Numbers*. As far as I am aware, *PopCo* is the first novel to interweave a romantic plot with recreational mathematics.

The narrator Alice is a young woman with an intense interest in play mathematics and the granddaughter of a mathematician who writes a column for a British newspaper similar to the one I wrote for *Scientific American*. My review of her novel ran in the *College Journal of Mathematics*, May 2007.

PopCo, a third novel by British author Scarlett Thomas, is as far as I know the only novel ever written that is saturated with a variety of topics in recreational mathematics. Alice Butler, the novel's narrator, twenty-nine years old, works for PopCo, the world's third-largest toy company. Her job is to create ideas for toys and games, especially products that will appeal to teenage girls.

Orphaned at an early age, Alice is raised by a loved grandfather and grandmother, both top mathematicians. The grandfather writes a monthly column on recreational math that is said to be similar to the one I wrote for *Scientific American*. He is obsessed with trying to decode the notorious Voynich manuscript, now widely believed to be a hoax. He fails in this project, but succeeds in decoding a cipher-text giving the location of a buried treasure. The treasure is on a bird sanctuary, and because digging it up would disturb the birds, the grandfather never reveals the treasure's location. He does, however, leave cipher clues to Alice. The grandmother's obsession is trying to solve the Riemann hypothesis.

Strongly influenced by her grandfather's fondness for mathematical play, Alice constantly interrupts her narrative with accurate discussions of topics that include the following: Conway's computer game of Life, Newcomb's famous paradox involving decision theory, trapdoor ciphers, and the Monty Hall problem of the three doors. It created an enormous flap when Marilyn vos Savant gave it in her *Parade* column. Although the problem involves only elementary probability, the result is so counterintuitive that thousands of mathematicians who should have known better accused Ms. Savant of ignorance when her solution was entirely accurate. Even the great Paul Erdös thought she was wrong until friend Ronald Graham explained it to him. The episode made the front page of the *New York Times*.

In addition to recreational topics, Alice also provides excellent accounts of Canto's alephs, how to factor primes, Gödel's proofs, even the Riemann hypothesis!

In addition to her knowledge of math, Alice is also a good player of both go and chess. She is addicted to cigarettes, smokes pot, plays the guitar, and has a cat named Atari. There is an amusing flashback to her high school days when she participated in a tournament run by a math teacher whom she despises. The winner gets to play the teacher. When he and Alice sit down to play he pompously tells her the game will probably be over in five minutes. Alice recognizes the teacher's opening as one used by Kasparov in a game she had once studied so carefully that she knows exactly how to meet the attack. Her grandfather had even written one of his columns about the game. Ten minutes later the teacher is defeated. He is so humiliated that he takes a long sick-leave.

The novel ends with two surprising climaxes. An expert in cipher breaking—one of Alice's PopCo toys is a code-breaking kit that includes Thomas Jefferson's cylinder with rotating wheels—Alice cracks her grandfather's cipher about the buried treasure. The treasure, worth billions, is unearthed, and Alice generously gives it all to the bird sanctuary.

The second climax is much more improbable. It concerns Alice's recruitment into a mysterious secret organization with a revolutionary agenda that I'll not reveal here in case a reader is moved to buy the book.

I was a bit put off by the frequency with which the book's characters, especially Ben, one of Alice's lovers, seem unable to speak without peppering their conversation with meaningless f-words. Do employees in British toy firms actually talk that way?

I was also surprised by Alice's low opinion of mainline doctors. I can understand how a woman as bright as she would be a vegan, a vegetarian who avoids all food coming from animals, including milk, cheese, eggs, and fish. Much harder to comprehend is why Alice is also a devotee of homeopathy. One wonders if this reflects Ms. Thomas's opinion, or only the views of her fictional narrator.

Surely Alice would know that homeopathic remedies dilute a drug so thoroughly that not even a molecule, or at most two or three, remain in the water or pill. Homeopaths are forced to assume that somehow the water "remembers" a drug's properties. Magician James Randi, the world's leading debunker of pseudoscience, in his lectures on homeopathy likes to flourish a homeopathic bottle in which a virulent poison has been diluted until only pure water remains. Randi then drinks the entire bottle.

The book includes a table for writing a Vigenére cipher. A postscript lists all primes less than 1,000, provides a clever crossword puzzle, and a recipe for baking a cake. Part 1 opens with a quotation from a paper by psychologist Stanley Milgram. Part 2 is preceded by a quote from Paul Hoffman's biography of Erdös, and Part 3 starts with one of Piet Hein's best "grooks":

A bit beyond perception's reach
  I sometimes believe I see
that life is two locked boxes, each
  containing the other's key.

Scarlett Thomas writes beautifully. Her novel is not easy to put down.

16.

FOUR LETTERS

I have an incurable habit of sending letters to newspapers and periodicals. It occurred to me that the following four recent letters might be amusing to readers of this book. The first letter, under the heading "What about Light, Dr. Sharp?" appeared in the *Norman* (Oklahoma) *Transcript*, October 1, 2005.

Editor, The Transcript:

I was surprised to see that you gave so much space to covering Dr. Thomas Sharp's lecture in which he argued that dinosaurs are mentioned 25 times in the Bible. He says the Old Testament calls them by the Hebrew word for dragon. The entire universe, according to Sharp, was created in six literal days, and on the sixth day, the same day that God created Adam and Eve, he created the dinosaurs. Presumably they all perished in the Great Flood because they were too big to fit on Noah's Ark.

Now, modern astronomy has overwhelming evidence that the universe is billions of years old. The best evidence is that billions of galaxies are so distant from Earth that it takes their light billions of years to reach us. I would be interested to know how Dr. Sharp accounts for this. Does he think light was created "on the way" about 10,000 years ago? Or does he think light traveled billions of times faster during the 6-day creation, then suddenly slowed down to its present speed of 186,000 miles per second?

Maybe your reporter could find out by asking Dr. Sharp which view he favors.

My second letter, which ran in the same paper on July 16, 2006, was headed "Ad Both Hilarious, Deplorable."

Editor, The Transcript:

A current TV commercial is as hilarious as it is deplorable. A pretty woman applies a small stick to her forehead while a voice-over repeats: "Head On, apply directly to the forehead."

Nothing is said about what the stick is supposed to do. Viewers are expected to deduce the stick must relieve headaches. The Food and Drug Administration can't take action because the ad makes no false claims. According to its box, the stick contains a homeopathic remedy.

I'm reminded of an ad said to have appeared in a newspaper decades ago. It said, "Please send me a dollar." When the flood of dollars dwindled, the ad was changed to "Last chance to send me a dollar."

Shame on Walgreens for selling this useless product. A friend in Florida tells me a clerk in a local Walgreens store said the sticks were selling like crazy. As Barnum put it, "A sucker is born every minute."

The ad ran for months on many TV stations around the country. It was usually followed by a second ad for an arthritis stick called Activon. "Applied directly to where it hurts," says the voice-over. Again there is no suggestion that the pain goes away or is even relieved.

Both sticks consist of a homeopathic salve. According to their containers, the drug has been diluted so many times that only one or two, perhaps no, molecules remain. The last time I saw the commercial it featured a black man who said he hated the commercial but the product was great. If you care to know the bizarre history of homeopathy, now experiencing a surprising revival in the United States and abroad, see the chapter on it in my Dover paperback, *Fads and Fallacies in the Name of Science*.

My third letter appeared in the Winter issue (no. 51, 2006) of John Wilcock's entertaining little magazine, *Ojai*, which he publishes from his home in Ojai, California. John tells me that he not only agrees completely with me, but that he often talks on his cable show about how unfunny *New Yorker* cartoons are, in contrast to cartoons in, say, the *Spectator*.

Editor, Ojai:

Why is it that *The New Yorker* cartoons are not funny anymore?

They used to be *very* funny. Back in the days when Harold Ross was editor the cartoons were uniformly amusing. I understand that Ross himself selected them. If so, he had a great sense of humor.

Is it possible I am merely feeling nostalgic for a lost youth? I don't think so. Recently I picked up some old *New Yorker*s and found the cartoons as hilarious as I remembered. There was Thurber's famous drawing of a man and woman in bed. She was saying "Whatever happened to the Socialist Party?" And there was a Steinberg cartoon showing parents peeking into a room where their son was supposed to be practicing on the violin. He was lying flat on his back, with the violin on the floor beside him. And there were great cartoons by Price that were always funny.

Today I occasionally find a *New Yorker* in a dentist's or doctor's office. I always check each cartoon. I have yet to find one that made me laugh. In many cases I'm even unable to discern what was supposed to be funny.

Cartoons in many foreign humor magazines are as funny as the old Ross cartoons. French and Russian, even British (though not *Punch*) humor magazines have very funny cartoons. I once subscribed to an Argentine magazine called *Juegos*. It was devoted entirely to puzzles, mathematical and linguistic. It also ran a few cartoons. I recall one that showed a husband returning home unexpectedly to find his wife in bed nude. A closet door was half open. A man inside was pointing toward the bed and saying to the husband, "Have you looked under the bed?" That broke me up.

But back to the great mystery. Ross liked to say his magazine was not edited for the old lady in Iowa who wears tennis shoes. I assume today's *New Yorker* is also not edited for such readers. Why then are its cartoons so dull, enigmatic, and unfunny?

Please, will someone explain.

The fourth letter ran in the *Norman Transcript*, October 29, 2006.

Editor, The Transcript:

Rush Limbaugh, on his radio show, is promoting a homeopathic remedy for the common cold. If Rush knew anything about the history and nature of homeopathy he would be much ashamed of this commercial. It is a basic dictum of this bogus medical cult that the greater a drug is diluted the more potent its effect. Accordingly, homeopathic drugs are diluted to the point where no molecules of the original substance remain. When you buy a homeopathic remedy you are buying only distilled water or a pure powder or salve. Homeopaths are forced to believe that the water or powder somehow "remembers" the properties of the original substance.

Reputable doctors tell a joke about a man who forgot to take his daily homeopathic pill and died of an overdose.

17.

IS BEAUTY TRUTH?

In the British edition of *Why Beauty Is Truth*, each chapter is headed by a stanza from Lewis Carroll's great nonsense ballad *The Hunting of the Snark*. All these stanzas were removed by Basic Books from the American edition of 2007. At the end of my review I have restored the stanza that has the most relevance to Stewart's book. The review appeared in the April 2007 issue of *Scientific American*.

Why Beauty Is Truth, the title of Ian Stewart's book (he has written more than sixty others) is, of course, taken from the enigmatic last two lines of John Keats's "Ode on a Grecian Urn":

> "Beauty is truth, truth beauty,"—that is all
>
> Ye know on earth, and all ye need to know.

But what on earth did Keats mean? T. S. Eliot called the lines "meaningless" and "a serious blemish on a beautiful poem." John Simon opened a movie review with "one of the greatest problems of art—perhaps the greatest—is that truth is not beauty, beauty not truth. Nor is it all we need to know." Stewart, a distinguished mathematician at the University of Warwick in England and a former author of *Scientific American* Mathematical Recreations column, is concerned with how Keats's lines apply to mathematics. "Euclid alone has looked on Beauty bare," Edna St. Vincent

Millay wrote. To mathematicians, great theorems and great proofs, such as Euclid's elegant proof of the infinity of primes, have about them what Bertrand Russell described as "a beauty cold and austere," akin to the beauty of great works of sculpture.

Stewart's first ten chapters, written in his usual easygoing style, constitute a veritable history of mathematics, with an emphasis on the concept of symmetry. When you perform an operation on a mathematical object, such that after the operation it looks the same, you have uncovered a symmetry. A simple operation is rotation. No matter how you turn a tennis ball, it does not alter the ball's appearance. It is said to have rotational symmetry. Capital "H" has 180-degree rotational symmetry because it is unchanged when turned upside down. It also has mirror reflection symmetry because it looks the same in a mirror. A swastika has 90-degree rotational symmetry but lacks mirror reflection symmetry because its mirror image whirls the other way.

Associated with every kind of symmetry is a "group." Stewart explains the group concept in a simple way by considering operations on an equilateral triangle. Rotate it 60 degrees in either direction, and it looks the same. Every operation has an "inverse" that cancels the operation. Imagine the corners of the triangle labeled *A*, *B*, and *C*. A 60-degree clockwise rotation alters the corners' positions. If this is followed by a similar rotation the other way, the original positions are restored. If you do nothing to the triangle, this is called the "identity" operation. The set of all symmetry transformations of the triangle constitutes its group.

Stewart's history begins with Babylonian and Greek mathematics, introducing their basic concepts in ways a junior high school student can understand. As his history proceeds, the math slowly becomes more technical, especially when he gets to complex numbers and their offspring, the quaternions and octonions. The history ends with the discoveries of Sophus Lie, for whom Lie groups are named, and the work of a little-known German mathematician, Joseph Killing, who classified Lie groups. Through this historical section, Stewart skillfully interweaves the math with colorful sketches of the lives of the mathematicians involved.

Not until the book's second half does Stewart turn to physics and explain how symmetry and group theory became necessary tools. A chapter on Albert Einstein is a wonderful blend of elementary relativity and details of Einstein's life. Next comes quantum mechanics and particle theory, with several pages on superstrings, the hottest topic in today's theoretical physics. Stewart is a bit skeptical of string theory, which sees all fundamental particles as inconceivably tiny filaments of vibrating energy that can be open-ended or closed like a rubber band. He does not mention two recent books (reviewed in the September 2006 issue of *Scientific American*) that give string theory a severe bashing. Lee Smolin's *The Trouble with Physics* denounces string theory as "not a theory at all," only a mishmash of bizarre speculations in search of a viable theory. Peter Woit's book is titled *Not Even Wrong*, a quote from the great Austrian physicist Wolfgang Pauli. He once described a theory as so bad it was "not even wrong."

Is string theory beautiful? Its promoters think so. Smolin and Woit believe that its recent absorption into a richer conjecture called M-theory has turned the former beauty of strings into mathematical structures as ugly as the epicycles Ptolemy invented to explain the orbits of planets as they circle the earth.

We are back to the mystery of Keats's notorious lines. In my opinion, John Simon is right. Even beautiful mathematical proofs can be wrong. In 1879 Sir Alfred Kempe published a proof of the four-color-map theorem. It was so elegant that for ten years it was accepted as sound. Alas, it was not. Henry Dudeney, England's great puzzle maker, published a much shorter and even prettier false proof.

In *The New Ambidextrous Universe* I write about the vortex theory of atoms. This popular nineteenth-century conjecture had an uncanny resemblance to superstrings. It maintained that atoms are not pointlike but are incredibly tiny loops of energy that vibrate at different frequencies. They are minute whirlpools in the ether, a rigid, frictionless substance then believed to permeate all space. The atoms have the structure of knots and links, their shapes and vibrations generating the properties of all the elements. Once created by the Almighty, they last forever.

In researching vortex theory, I came across many statements by eminent physicists, including Lord Kelvin and James Clerk Maxwell, suggesting that vortex theory was far too beautiful not to be true. Papers on the topic proliferated, and books about it were published. Scottish mathematician Peter Tait's work on vortex atoms led to advances in knot theory. Tait predicted it would take several generations to develop the theory's mathematical foundations. Beautiful though it seemed, the vortex theory proved to be a glorious road that led nowhere.

Stewart concludes his book with two maxims. The first: "In physics, beauty does not automatically ensure truth, but it helps." The second: "In mathematics beauty *must* be true—because anything false is ugly." I agree with the first statement, but not the second. We have seen how lovely proofs by Kempe and Dudeney were flawed. Moreover, there are simply stated theorems for which ugly proofs may be the only ones possible.

Let me cite two recent examples. Proof of the four-color-map theorem required a computer printout so vast and dense that it could be checked only by other computer programs. Although there may be a beautiful proof recorded in what Paul Erdös called "God's book"—a book that, he suggested, included all the theorems of mathematics and their most beautiful proofs—it is possible that God's book contains no such proof. The same goes for Andrew Wiles's proof of Fermat's last theorem. It is not computer-based, but it is much too long and complicated to be called beautiful. There may *be* no beautiful proof for this theorem. Of course, mathematicians can always hope and believe otherwise.

Because symmetry is the glue and tape that binds the pages of Stewart's admirable history, a stanza from Lewis Carroll's immortal nonsense ballad *The Hunting of the Snark* could serve as an epigraph for the book:

> You boil it in sawdust: you salt it in glue:
> You condense it with locusts and tape:
> Still keeping one principal object in view—
> To preserve its symmetrical shape.

18.

IS STRING THEORY IN TROUBLE?

I can't pretend that I understand the advanced mathematics of string/M-theory. I can only say, as Einstein once remarked about quantum mechanics, that my little finger tells me that string theory is at least incomplete, if not totally wrong. Of course, I could be mistaken. My review of Lee Smolin's *The Trouble with Physics* (Houghton Mifflin, 2007), and Peter Woit's *Not Even Wrong* (Basic Books, 2007) ran in *New Criterion*, April 2007.

For more than thirty years, string theory has been what Murray Gell-Mann called "the only game in town." By this he meant that it was the only good candidate for a TOE, or Theory of Everything. Not only does it claim to unify relativity and quantum mechanics; it also explains the existence of all fundamental particles. Instead of being "pointlike," they are modeled by filaments of energy so tiny that there is no known way to observe them or even to prove they are real.

A string can have two ends or be closed like a rubber band. Of great tensile strength, strings vibrate at different frequencies. They live in a space of ten or eleven dimensions, of which six or seven are "compacted" into inconceivably minute structures attached to every point in our four-dimensional space-time. The simplest vibration of a closed string produces a graviton, the quantized particle of gravity. One of string theory's earliest triumphs was forcing the reality of gravitons.

After an obscure, bumbling start, string theory slowly began to gain momentum until it became the hottest topic in physics. Thousands of papers were published and thick textbooks written. The fastest way to advance in departments of great universities was to work on strings. Richard Feynman and Sheldon Glashow were almost alone among famous physicists who were skeptical of the trend. Not until a few years ago did skepticism begin to surge. Simmering doubts reached a boiling point last September when two eminent physicists published slashing attacks on string theory. Their books may mark a dramatic turning point in the history of modern physics.

For years, Lee Smolin rode the string bandwagon. After teaching at Yale and Penn State, he became a researcher at the Institute for Theoretical Physics in Waterloo, Canada, a think tank he helped found. *The Trouble with Physics*, his third book, is a powerful indictment. He sees string theory as not a theory—only a set of curious conjectures in search of a theory. True, it has great explanatory power, but a viable theory must have more than that. It must make predictions which can be falsified or confirmed.

In addition to this whopping lack of evidence, string theory has suffered other setbacks. It has been absorbed into a richer set of conjectures called M-theory. The M stands mainly for membranes (branes for short), or for Magic, Mystery, Mother of all theories, or any other term you like that begins with M. In M-theory, strings are one-dimensional branes that can roam free or be attached to two-dimensional branes. Branes may be of any dimension from 1 through 9. One wild speculation is that our 3-brane universe floats within a monstrous higher-dimension brane. To a mere science journalist like myself, the great mathematical beauty of early string theory has degenerated into M for Messy. Its membranes, in Smolin's opinion, are as ugly as the epicycles Ptolemy fabricated to describe the curious paths of planets as they seem to circle Earth.

The most troubling aspect of string/M-theory is that the compacted dimensions, known as Calabi-Yau manifolds, can take at least a hundred thousand different shapes. This has led to the mind-boggling concept of a vast "landscape" consisting of a multiverse containing a hundred thousand, perhaps an infinity, of universes, each with its own Calabi-Yau space! Every universe would have a random selection of physical constants, such as the velocity of light. By anthropic reasoning, we of course live in a universe with just the right set of constants that make possible galaxies, stars, planets, and, on one small planet, such bizarre creatures as you and me.

Other string/M-theory embarrassments are carefully detailed by Smolin. Cosmologists have discovered that most of our universe consists of "dark matter," so called because it is totally invisible. String theorists failed to predict it, and have nothing useful to say about it. A more recent discovery is that the universe is expanding at a slightly increasing rate. Such acceleration can only be caused by the pressure of a mysterious "dark force." Again, writes Smolin, dark force was not predicted by string theory, and the theory has no good explanation for it.

A chapter in Smolin's persuasive book divides physicists into two classes: craftsmen who test theories; and seers, like Newton and Einstein, who create theories. What physics now desperately needs, Smolin is convinced, is a new Einstein who can replace M-theory with a TOE that can be confirmed by a workable experiment.

Another chapter is devoted to lonely seers, working patiently outside the establishment on conjectures as revolutionary as string theory. Roger Penrose, Oxford's famous mathematical physicist, is the best-known seer. His twistor theory, alas also untestable, is M-theory's chief rival. Like many other seers, Penrose thinks Einstein was right to regard quantum mechanics as "incomplete." Other intrepid seers are starting to question even special relativity. Because both relativity and quantum mechanics are essential to M-theory, finding either theory in need of revision would be, Smolin writes, another severe blow to string/M-theory.

In a chapter on sociology, Smolin introduces the concept of "groupthink"—the tendency of groups to share an ideology. This creates a cult-like atmosphere in which those who disagree with the ideology are considered ignoramuses or fools. Most physicists tied up in the string mania, Smolin believes, have become groupthinkers, blind to the possibility that they have squandered time and energy on weird speculations that are leading nowhere.

Figure 17: String theory cartoon. © The New Yorker Collection 2007 Lee Lorentz from cartoonbank.com. All rights reserved.

In spite of such criticisms Smolin, like Edward Witten, by far the most energetic and creative of the stringers, believes that even if string/M-theory is finally abandoned, portions of it will remain fruitful. Peter Woit, a mathematical physicist at Columbia University, is less optimistic. He sees little hope that any aspect of M-theory will survive. The harshness of his rhetoric is signaled by his book's arresting title, *Not Even Wrong*. It's a famous quote from the great Austrian physicist Wolfgang Pauli. A certain theory was so bad, he said, that "it was not even wrong." By this he meant it was so flimsy it couldn't be confirmed or falsified.

Most of Woit's book is a moderately technical, equation-free survey of quantum mechanics, the standard model of particle theory, and the history of superstrings. The prefix "super-" indicates the linkage of strings to an earlier theory called supersymmetry. Not until the last third of his book does Woit take up reasons for regarding string theory a failure, destined to give way to a testable TOE.

Although Woit sees Edward Witten as the guru of what resembles a religious cult, he has only the highest respect for Witten's genius. Amazingly, Witten's early training was in economics. He soon shifted to mathematics and physics at Princeton University. There, he obtained his doctorate and became a professor for several years before moving to New Jersey's Institute for Advanced Study, where he has remained ever since. He has been given a MacArthur Award, and a Fields Medal, the mathematical equivalent of a Nobel Prize.

When Woit was a graduate student at Princeton, he once followed Witten up a stairway from a library to a plaza. When he reached the plaza, Witten had mysteriously vanished. "It crossed my mind," Woit writes, "that a consistent explanation . . . was that Witten was an extraterrestrial being from a superior race who, when he thought no one was looking, had teleported back to his office."

Woit's main objection to string theory, of course, is that it has not, in Glashow's words, "made even one teeny-tiny experimental prediction." Woit quotes Feynman: "String theorists do not make predictions, they make excuses."

In his book *Interactions*, Glashow writes:

> Until string people can interpret perceived properties of the real world they simply are not doing physics. Should they be paid by universities and be permitted to pervert impressionable students? Will young Ph.D's, whose expertise is limited to superstring theory, be employable if, and when, the string snaps? Are string thoughts more appropriate to departments of mathematics, or even to schools of divinity, than to physics departments? How many angels can dance on the head of a pin? How many dimensions are there in a compacted manifold, 30 powers of ten smaller than a pinhead?

Woit quotes from another Nobel Prize winner, the Dutch physicist Gerard 't Hooft:

> Actually, I would not even be prepared to call string theory a "theory" rather a "model" or not even that: just a hunch. After all, a theory should come together with instructions on how to deal with it to identify the things one wishes to describe, in our case the elementary particles, and one should, at least in principle, be able to formulate the rules for calculating the properties of these particles, and how to make new predictions for them. Imagine that I give you a chair, while explaining that the legs are still missing, and that the seat, back and armrest will perhaps be delivered soon; whatever I did give you, can I still call it a chair?

Woit has only harsh things to say about the recent acceptance of an anthropic principle by several leading string theorists, notably Steven Weinberg and David Susskind. Susskind has even written a popular book about it—*The Cosmic Landscape: String Theory and the Illusion of Intelligent Design*. The notion that there could be millions of other universes, each with its own Calabi-Yau structure—or what amount to the same thing, with its own basic state of what physicists like to call the "vacuum"—is not one that appeals to Witten. "I'd be happy if it is not right," Woit quotes from a 2004 lecture, "but there are serious arguments for it, and I don't have any serious argument against it."

In the nineteenth century, a conjecture called the vortex theory of the atom became extremely popular in England and America. Proposed by the famous British physicist Lord Kelvin, it had an uncanny resemblance to string theory. It was widely believed at the time that space was permeated by an incompressible frictionless fluid called the ether. Atoms, Kelvin suggested, are super-small whirlpools of ether, vaguely similar to smoke rings. They take the form of knots and links. Point particles can't vibrate. Ether rings can. Their shapes and frequencies determine all the properties of the elements. Vortex theory isn't mentioned by Woit, although Smolin considers it briefly.

Kelvin published two books defending his conjecture. It was strongly championed in England by J. J. Thompson in his 1907 book, *The Corpuscular Theory of Matter*. Another booster of the theory was Peter Tait, an Irish mathematician. His work, like Witten's, led to significant advances in knot theory. In the United States, Albert Michelson considered vortex theory so "grand" that "it ought to be true even if it is not." Hundreds of papers elaborated the theory. Tait predicted it would take generations to develop its elegant mathematics. Alas, beautiful though vortex theory was, it proved to be a glorious road that led nowhere.

Will string theory soon meet a similar fate? Glashow wrote a clever poem that he recited at a Grand Unification Workshop in Japan. It ends with the following lines:

> Please heed our advice that you too are not smitten—
> The book is not finished, the last word is not Witten.

19.

DO LOOPS EXPLAIN CONSCIOUSNESS?

When Douglas Hofstadter's great book *Gödel, Escher, Bach* was published I devoted my *Scientific American* column to a rave review. Since then Doug and I have become friends. His book *I Am a Strange Loop* (Basic Books, 2007) is as delightfully written as his earlier books, and swarming with his familiar clever wordplay. I was reluctant at first to review *I Am a Strange Loop* because I disagree with his main theme, but finally I decided I might as well defend my point of view. The review appeared in the August 2007 issue of *Notices of the AMS*. For an earlier essay on the topic see chapter 36, "Are Computers Near the Threshold?" in my *Night Is Large*. For my views on free will see chapter 6 of *Whys of a Philosophical Scrivener*.

Barmaid: "Would you like some wine?"
Descartes: "I think not."
Then he vanishes.
—*Anonymous joke*

Our brain is a small lump of organic molecules. It contains some hundred billion neurons, each more complex than a galaxy. They are connected in over a million billion ways. By what incredible hocus-pocus does this tangle of twisted filaments become aware of itself as a living thing, capable of love and hate, of writing novels and symphonies, feeling pleasure and pain, with a will free to do good and evil?

David Chalmers, an Australian philosopher, has called the problem of explaining consciousness the "hard problem." The easy problem is understanding unconscious behavior, such as breathing, digestion, walking, perceiving, and a thousand other things. Grappling with the hard problem has become one of the hottest topics facing philosophers, psychologists, and neuroscientists. According to philosopher John Searle, reviewing Nicholas Humphrey's *Red: A Study of Consciousness* (*New York Review of Books*, November 2005), Amazon lists 3,865 books on consciousness. The most recent, published this year by Basic Books, is Douglas Hofstadter's *I Am a Strange Loop*.

Hofstadter, a professor of cognitive science at Indiana University, is best known for his Pulitzer Prize–winning *Gödel, Escher, Bach*, or *GEB* as he likes to call it. His new book, as brilliant and provocative as earlier ones, is a colorful mix of speculations with passages of autobiography. An entire chapter is devoted to a terrible tragedy that Hofstadter is still trying to cope with. His wife Carol, at age forty-two, died suddenly of a brain tumor. The preceding chapter links his love for Carol to a fantasy he once conceived about a mythical land he called Twinwirld. Its inhabitants are identical twins, so nearly alike that they think and act like single individuals.

I suspect Hofstadter will be surprised to know that L. Frank Baum, in his non-Oz fantasy *The Enchanted Island of Yew* (you?), imagined a similar land he called Twi. Everything in Twi is duplicated, like seeing the world through glasses that produce double images. Residents of Twi, like those of Twinwirld, are identical twins. The rulers of Twi are two beautiful girls who think and speak as a single entity.

In his heartrending chapter on Carol, Hofstadter makes clear why he preceded it with a description of Twinwirld. He and Carol were so much alike they resembled a pair of Twinwirlders. Unable to find consolation in hope for an afterlife, Hofstadter's only solace is knowing that for at least a time Carol will in a way live on in the memories of those who knew and loved her.

I Am a Strange Loop swarms with happy memories. One vivid recollection, not so happy, concerns a time when Hofstadter was fifteen and asked to select two guinea pigs to be killed for a laboratory experiment. Faced with the task, he fainted. This aversion to animal killing led to his becoming a vegetarian. For a while he allowed himself eggs and fish, but later became a vegan, avoiding all food of animal origin. He refuses to buy leather shoes and belts. Like Baum's Tin Woodman, to whom the Wizard gave a fine velvet heart, Hofstadter has twinges of guilt when he swats a fly. One of his heroes is Albert Schweitzer, who whenever possible avoided killing an insect.

Many pages in *I Am a Strange Loop* express the author's great love of music. Hofstadter plays a classical piano. Bach, Chopin, and Prokofiev are among his favorite composers, Bartok among those he dislikes. Another passion is for poetry. He has translated from the Russian Pushkin's great poem *Eugene Onegin*, as well as the work of other foreign bards.

On page 94 Hofstadter offers a clever six-stanza poem by a friend that commemorates an event he later considered symbolic. One day he grabbed a batch of empty envelopes and was puzzled by what seemed to be a marble wedged between them. The marble turned out to be a spot where a thickness of paper felt like a marble. In a similar way, he believes, we imagine a self wedged somewhere between the neurons of our brain.

The marble provides the central theme of *I Am a Strange Loop*. The soul, the self, the I, is an illusion. It is a strange loop generated by a myriad of lesser loops. It is a minute portion of the universe, a glob of dead matter within our skull, not only observing itself, but aware it is observing itself.

Hofstadter has long been fascinated by self-reference loops. He sees them everywhere. They are at the heart of Gödel's famous undecidability proof. They lurk within Russell and Whitehead's *Principia Mathematica*. They are modeled by such logic paradoxes as "This sentence is false" and by the card that says on one side "The sentence on the other side is true," and on the flip side says "The sentence on the other side is false." Similar loops are such lowly feedback mechanisms as flywheels, thermostats, and flush toilets. He reproduces Escher's famous lithograph of two hands, each drawing the other, and suggests modifying it by having one hand erase the other.

Many photographs in the book depict recursive loops. One shows a carton closed by four flaps, A on top of B, B on top of C, C on D, and D on A. In another picture Doug and Carol are each touching the other's nose. An amusing photo shows a grinning Doug with nine friends, each sitting on the lap of a person behind.

In chapter 21 Hofstadter introduces a disturbing thought experiment, involving human identity, that has been central in dozens of science fiction tales. A man is teleported by a process made famous by *Star Trek.* Officers of the *Enterprise* are beamed down to a planet, and later beamed up again. This is done by apparatus that scans a person molecule by molecule, then transmits the information to a distant spot where it creates an exact duplicate of the person. If this destroys the original body there is no philosophical difficulty. But suppose the original is not destroyed. The result is a pair of identical twins with identical memories. Is the teleported person the *same* person or someone else?

The dilemma goes back to Plutarch. He imagines a ship that is slowly replaced, piece by piece, until the entire ship is reconstituted. The original parts are then reassembled. Each ship can claim to be the original.

Baum introduces the same problem in his history of the Tin Woodman. As all Oz buffs know, a cruel witch enchants Nick Chopper's ax, causing it to slice off parts of Nick's meat body. Each part is replaced by Ku-Klip, a master tinsmith, until Nick is made entirely of tin. In *The Tin Woodman of Oz* the tin man visits Ku-Klip's workshop where he converses with his former head. Ku-Klip has preserved it in a cupboard. Who is the real Nick Chopper? The tin man or his former head?

Hofstadter has little interest in such conundrums. Another topic that infuriates him is free will. Unlike his good friend philosopher Daniel Dennet, Hofstadter denies that free will exists. It is another mirage, like the marble in the envelopes.

Other topics drive Hofstadter up a wall. One is the "inverted spectrum" parardox. How can we be certain that our sensation of, say, red is the same as that of another person? What we experience as red could be what she experiences as what we call blue.

Another topic Hofstadter considers frivolous is the concept of a zombie. Zombies are persons who think, talk, and behave exactly like ordinary people but are entirely lacking in all human feelings and emotions. The concept arises in relation to computerized robots. Baum's wind-up robot Tik-Tok, who Dorothy rescues in *Ozma of Oz*, has a metal plate on his back that says, "Thinks, Speaks, Acts, and Does Everything but Live." It is hard to believe, but entire books have been written about zombies.

Consciousness for Hofstadter is an illusion, along with free will, although both are unavoidable, powerful mirages. We feel as if a self is hiding inside our skull, but it is an illusion made up of millions of little loops. In a footnote on page 374 he likens the soul to a "swarm of colored butterflies fluttering in an orchard."

Like his friend Dennet, who wrote a book brazenly called *Consciousness Explained*, Hofstadter believes that he too has explained it. Alas, like Dennet, he has merely described it. It is easy to describe a rainbow. It is not so easy to explain a rainbow. It is easy to describe consciousness. It is not so easy to explain the magic by which a batch of molecules produce it. To quote a quip by Alfred North Whitehead, Hofstadter and Dennet "leave the darkness of the subject unobscured."

Let me spread my cards on the table. I belong to a small group of thinkers called the "mysterians." It includes such philosophers as Searle (he is the scoundrel of Hofstadter's book), Thomas Nagel, Colin McGinn, Jerry Fodor, also Noam Chomsky, Roger Penrose, and a few others.

We share a conviction that no philosopher or scientist living today has the foggiest notion of how consciousness, and its inseparable companion free will, emerge, as they surely do, from a material brain. It is impossible to imagine being aware we exist without having some free will, if only the ability to blink yes or no, or to decide what to think about next. It is equally impossible to imagine having free will without being at least partly conscious.

In dreams one is dimly conscious but usually without free will. Vivid out-of-body dreams are exceptions. Many decades ago, when I was for a short time taking tranquilizers, I was fully aware in out-of-body dreams that I was dreaming, but could make genuine decisions. In one dream, when I was in a strange house, I wondered if I could produce a loud noise. I picked up a heavy object and flung it against a mirror. The glass shattered with a crash that woke me. In another OOB dream I lifted a burning cigar from an ashtray, and held it to my nose to see if I could smell it. I could.

We mysterians are persuaded that no computer of the sort we know how to build—that is, one made with wires and switches—will ever cross a threshold to become aware of what it is doing. No chess program, however advanced, will know it is playing chess anymore than a washing machine knows it is washing clothes. Today's most powerful computers differ from an abacus only in their power to obey more complicated algorithms, to twiddle ones and zeroes at incredible speeds.

A few mysterians believe that science, some glorious day, will discover the secret of consciousness. Penrose, for example, thinks the mystery may yield to a deeper understanding of quantum mechanics. I belong to a more radical wing. We believe it is the height of hubris to suppose that evolution has stopped improving brains. Although our DNA is almost identical to a chimpanzee's, there is no way to teach calculus to a chimp, or even to make it understand the square root of 2. Surely there are truths as far beyond our grasp as our grasp is beyond that of a cow.

Why is our universe mathematically structured? Why does it, as Hawking recently put it, bother to exist? Why is there something rather than nothing? How do the butterflies in our brain—or should I say bats in our belfry—manage to produce the strange loops of consciousness?

There may be advanced life-forms in Andromeda who know the answers. I sure don't. Nor do Hofstadter and Dennet. And neither do you.

PART TWO: LITERATURE

20.

CHESTERTON: *THE FLYING INN*

I sometimes imagine that I am the only non-Catholic who is an enthusiastic admirer of Gilbert Chesterton. I'm well aware of his few faults—his unconscious anti-Semitism, his ignorance of science, his naïve political views—but I share his faith in a personal God, above all in his celebrated emotions of awe, wonder, and gratitude for the existence of a universe (why should there be anything rather than nothing?) and the even more miraculous existence of himself. Moreover, I enjoy his robust humor, his swashbuckling style, his poetry, and the ingenuity of his fiction.

G. K.'s novel *The Flying Inn* has a timely interest because it concerns the threat of an Islamic takeover of the British government. My essay appeared in *New Criterion* (March 2006). For more of M. G. on G. K. see my *Annotated Man Who Was Thursday*, my *Annotated Innocence of Father Brown*, and my introductions to five Dover paperback editions of other fictional works by Chesterton.

Like Gilbert Chesterton's *The Ball and the Cross*, *The Flying Inn* is a comic fantasy almost totally forgotten today, even by Chestertonians. In view of the current explosion of Islamic fundamentalism, and the rise of terrorism against infidel nations, *The Flying Inn* has an eerie relevance to the Iraq war that keeps the novel from flying into complete obscurity.

Lord Phillip Ivywood, the novel's main character, is England's handsome, golden-voiced prime minister. He has come under the influence of a Turkish fanatic, Misyara Ammon, a large-nosed, black-bearded Muslim popularly known as the Prophet of the Moon. He has convinced Lord Ivywood that the Muslim faith is superior to Christianity. It is a progressive force destined to dominate the world. Ivywood has decided that Christianity and Islam should merge, with the Muslim crescent placed alongside the cross on top of London's St. Paul's Cathedral. Better yet, the cross should be abandoned for a new symbol that combines cross and crescent, perhaps called the "crosslam."

The universities of England are filled with Muhammadan students. Soon they will be a majority. Ivywood orders that the curved line of a crescent moon replace the cross on all voting ballots. A children's game similar to tic-tac-toe is in all toy stores. It is called "naughts and crescents."

Swayed by the passionate rhetoric of the Turkish mystic, Lord Ivywood has become a strict vegetarian. In an effort to abolish all alcoholic drinks, Ivywood closes down inns that sell booze. His enormous home near the seaside fishing village of Pebblewick is lavishly redecorated in Arabian style. In the interests of world peace, he has allowed the British army to dwindle until it is a token force. The so-called Higher Criticism of biblical miracles has become fashionable throughout England, thanks to the learned writings of theologians with such names as Widge, Gilp, Bunk, Poote, Minns, and Toscher.

Patrick Dalroy is a huge red-haired Christian, a former sea captain from Ireland, presumably a Catholic. He has long been in love with Lady Joan Brett, a distant cousin of Lord Ivywood. She lives as a guest in his Pebblewick mansion.

Dalroy retires to Pebblewick. His friend Humphrey Pump owns a tavern in Pebblewick called The Old Ship. When the inn is closed down by Ivywood's orders, Humphrey and Dalroy buy a cart and donkey, then turn the cart into a "flying inn" which they take here and there to supply rum and cheese to nearby residents.

I'll not attempt to go into the novel's intricate, bewildering plots and subplots. There are many hilarious episodes. Dorian Wimpole, another cousin of Ivywood, is a poet entranced by all living creatures. He is known as the Poet of the Birds because of his popular volume of verse about songbirds. One of his sonnets is titled "Motherhood" because it is about the mother of a scorpion. Other poems concern walruses. A chapter covers his speech on the glory of oysters.

Another comic episode ridicules an exhibit of Post-Futurist art. A rustic who denies he is intoxicated is offered another drink if he can study two paintings, one of an old woman, the other said to depict rain in the hills, and tell which is which. Unable to decide, he mumbles, "I must be drunk after all."

One of the novel's most believable—certainly the most lovable—characters is a mongrel dog named Quoodle. Chapter 10 is devoted almost entirely to Quoodle.

Lady Joan is a great admirer of Lord Ivywood, although she is becoming more and more disturbed by his growing devotion to Islam. The endless rooms and corridors of his mansion, gaudily decorated like an Arabian palace, begin to suggest to her a harem. Her intuitions tell her that Ivywood is moving toward legitimizing polygamy throughout England. Near the book's end, Ivywood proposes marriage. Joan asks for time to think it over.

The novel comes to a wild climax when Captain Dalroy mobilizes an army of revolt that marches on Ivywood's estate. It is met by a defending army led by Omar Pasha, a Turk with a big scar on his face. He and Dalroy actually clash swords, an event symbolic of the conflict between the two great religions. The Captain is slightly cut on his wrist and forehead, but Pasha is mortally wounded.

On the book's last pages we learn that Ivywood, now mentally deranged but happy, is being cared for by his first cousin Lady Enid Wimpole, a relative of the poet Dorian Wimpole.

Dorian is not the only poet in the novel. Captain Dalroy and Humphrey are also poets who periodically recite or sing their verse. The only poem worth a second reading is "The Song of Quoodle." The dog reflects on the poor human sense of smell:

And Quoodle here discloses
All things that Quoodle can;
They haven't got no noses,
They haven't got no noses,
And goodness only knowses
The noselessness of Man.

Lady Joan and Captain Dalroy, I'm pleased to report, are finally married. At least, this is implied when Joan tells her cousin Enid, "We are so happy." To which Enid responds, "But his happiness will last." It's a puzzling remark. Is Enid hinting that maybe Joan's happiness will *not* last?

The Flying Inn seems to me Chesterton's least successful work of fiction. Its paragraphs tend to be long and convoluted, the dialogue dubious, and the narrative too often interrupted by tiresome poems and song lyrics. For admirers of G. K., however, it is still worth reading for its interesting ideas, its beautifully worded sentences, and its gorgeous sunsets.

In rereading the novel I found myself thinking of a haunting story by the Irish writer Lord Dunsany in his *Last Book of Wonder*. Dunsany visits an old, decaying castle in Provence, France. There he encounters the spirit of the tower, an old man with a long white beard and carrying a horn. "Beware, beware," he says.

"Beware of what?" Dunsany asks.

"The Saracens," the old man answers. "And I shall blow my horn."

> Then I explained, so that he might have rest, and told him how all Europe, and in particular France, had terrible engines of war, both on land and sea; and how the Saracens had not these terrible engines either on sea or land, and so could by no means cross the Mediterranean or escape destruction on shore even though they should come there. I alluded to the European railways that could move faster than horses could gallop. And when as well as I could I had explained all, he answered, "In time all these things pass away and then there will still be the Saracens."
>
> And then I said, "There has not been a Saracen either in France or Spain for over four hundred years."
>
> And he said, "The Saracens! You do not know their cunning. That was ever the way of the Saracens. They do not come for a while, no not they, for a long while, and then one day they come."

> And peering southwards, but not seeing clearly because of the rising mist, he silently moved to his tower and up its broken steps.

Well, the Saracens are indeed coming! England and America are not likely to have a prime minister or president who admires Islam, but Lord Ivywood's love of Islam is not much different form America's growing cultural relativism and political correctness. Our recent flap over the word "Christmas" is not much different from Lord Ivywood's efforts to get rid of the cross. How the war against Islamic terrorism will end, who knows? We can only hope and pray it won't end in a nuclear duel.

The Flying Inn was first published in England in 1914 by Methuen. The American edition appeared the same year by John Lane. The book's songs and poems had earlier appeared in issues of the *New Witness*, and later were collected in *Wine, Water, and Song* (Methuen, 1915).

The Flying Inn, together with two other novels, are reprinted in volume VII of the Ignatius Press series of Chesterton's *Collected Works*. The editor, Ian T. Benson, provides an introduction and a wealth of valuable footnotes. Volume XI of the collected works includes a musical stage version of *The Flying Inn* that G. K. wrote years before he wrote the novel. Virtually unknown to Chestertonians, the musical includes some characters (for example, the Duchess of Battle-axe) who are not in the novel.

Reviews of the novel were mostly unfavorable. Hubert Bland, writing in the *New Statesman* (January 21, 1913), had this to say:

> Mr. Chesterton appears to be just tumbling about, as it were, rollicking, roaring, sometimes spluttering half-articulate maledictions, sometimes bellowing resonant guffaws, always trusting to inspiration, and, for the most part, trusting in vain. . . . I was on the point of letting the book drop from my hands in sheer boredom, when there came a bright and penetrating flash of wit or a lurid gleam of humour which bucked me up and set me reading on once more.

21.

CHESTERTON: *MANALIVE*

Chesterton's improbable novel *Manalive* is seldom read today, but in my opinion it is one of his finest works of fantastic fiction. My tribute to it is scheduled to appear in *Gilbert Magazine*, a periodical devoted to Chesterton.

Among Chesterton's full-length novels, the greatest of course is *The Man Who Was Thursday*. In my opinion his second-most memorable novel is *Manalive*. Although the story is high comedy, it is woven around what G. K. called the central idea of his life: a belief that one should see everything, from the universe down to a daisy, with an emotion of awe and wonder combined with gratitude for the amazing privilege of being alive.

Why, Stephen Hawking has famously asked, does the universe bother to exist? Paul Edwards, in his *Encyclopedia of Philosophy*, calls this the "superultimate question." Obviously there is no way to answer, but the question is not therefore meaningless. Indeed, reflecting on it can arouse an emotion so close to terror that if it lasted more than a few moments one might go mad. On a humbler level the emotion is one of surprise toward ordinary things, what G. K. called in the title of one of his best collections of essays *Tremendous Trifles*. Some of them are mentioned in one of his finest poems, "A Second Childhood." Allow me to quote it in full:

When all my days are ending
And I have no song to sing,

I think I shall not be too old
To stare at everything;
As I stared once at a nursery door
Or a tall tree and a swing.

Wherein God's ponderous mercy hangs
On all my sins and me,
Because He does not take away
The terror from the tree
And stones still shine along the road
That are and cannot be.

Men grow too old for love, my love,
Men grow too old for wine,
But I shall not grow too old to see
Unearthly daylight shine,
Changing my chamber's dust to snow
Till I doubt if it be mine.

Behold, the crowning mercies melt,
The first surprises stay;
And in my dross is dropped a gift
For which I dare not pray:
That a man grow used to grief and joy
But not to night and day.

Men grow too old for love, my love,
Men grow too old for lies;
But I shall not grow too old to see
Enormous night arise,
A cloud that is larger than the world
And a monster made of eyes.

Nor am I worthy to unloose
The latchet of my shoe;
Or shake the dust from off my feet

Or the staff that bears me through
On ground that is too good to last,
Too solid to be true.

Men grow too old to woo, my love,
Men grow too old to wed;
But I shall not grow too old to see
Hung crazily overhead
Incredible rafters when I wake
And find I am not dead.

A thrill of thunder in my hair:
Though blackening clouds be plain,
Still I am stung and startled
By the first drop of the rain:
Romance and pride and passion pass
And these are what remain.

Strange crawling carpets of the grass,
Wide windows of the sky:
So in this perilous grace of God
With all my sins go I:
And things grow new though I grow old,
Though I grow old and die.

Manalive is about a man, Innocent Smith, who has a twofold mission: to invent curious ways, close to practical jokes, of keeping himself in a perpetual state of Chestertonian wonder, and to arouse a similar state in others. The crazy novel opens with a great alliterative sentence: "A wind sprang high in the west, like a wave of unreasonable happiness." The wind is a symbol of the wave of wonder that pervades all of Smith's slapstick antics.

Beacon House is a London establishment owned by Mrs. Drake. Among the residents are her niece Diana, Rosamund Hunt, an Irish journalist named Michael Moon, Arthur Inglewood, and Mary Gray, a newcomer. Miss Gray, the book's most intriguing character, is a shy young woman with red hair. She seldom speaks, but seems almost about to speak.

Inglewood receives a cryptic telegram from his old friend Innocent Smith. It reads: "Man found alive with two legs."[1] Smith himself arrives on the scene by leaping over a garden wall. He is a giant of a man, with blond hair, a large pointed nose, and a head that seems small in comparison to his apelike body.

Figure 18: Frontispiece of the American edition of *Manalive*.

When the strong wind blows off Smith's hat, Inglewood starts to run after it, but Smith stops him by shouting "Unsportsmanlike!! Give it fair play!" Smith chases his hat, finally falling on his back and catching the hat between his shoes. The wind then blows off the hat of Dr. Herbert Werner, a physician visiting Beacon House. His hat, caught in a tree, is retrieved by Smith, who climbs the tree like an acrobat.[2]

Smith asks Mrs. Drake for lodging. She takes him to a room in the attic, and in no time Smith has found a trapdoor in the slated ceiling, and is on the roof inviting persons up for a picnic.

Smith unpacks his yellow Gladstone bag. On it are the initials I. S. It is no coincidence that they spell "is"—the *est* of Thomas Aquinas. To invoke President Bill Clinton, "is" is a synonym for "to be." Among Smith's things are an odd assortment of trivial objects, such as six small wine bottles which he selected because their labels have the six colors of a spectrum.

Among his luggage is a revolver. Asked if he carries it for purposes of killing, he replies "I deal life out of that."

Soon Smith's enthusiasm for commonplace objects is infecting everyone at Beacon House. He uses colored chalk to create patterns on Diana's drab clothes. "All is gold that glitters," he says—an aphorism that could have come straight out of Veblen's *Theory of the Leisure Class*. Smith follows the remark with a jingle he apparently wrote:

> All is gold that glitters—
> Tree and tower of brass;
> Rolls the golden evening air
> Down the golden grass.
> Kick the cry to Jericho,
> How yellow mud is sold;
> All is gold that glitters,
> For the glitter is the gold.

An event now occurs that shocks all the boarders. Smith takes Mary Gray into the garden to observe one of G. K.'s gorgeous sunsets, and there he proposes marriage! Incredibly, Mary accepts! (Two other romances run through the novel. They involve Moon and Rosamund, Inglewood and Diana.) Rosamund, thinking Smith has gone insane, sends a telegram of help to Dr. Werner.

Werner arrives at Beacon House bringing with him Dr. Cyrus Pym, an American physician who is also a private detective. They arrest Smith, with the intention of taking him first to Scotland Yard then to an asylum. Dr. Pym has checked police records and learned that Smith, under various names, has married and abandoned a variety of young women and possibly murdered them. He is also wanted for burglary, and for attempting to shoot a Cambridge University professor of philosophy.

When Mary Gray is told about Smith's crimes, she astounds everyone by saying she knows all about them but intends to marry Smith anyway. Her only comment about his crimes is "Why, that must be awfully exciting." When told a burglary charge was signed by an Anglican curate, she responds, "Oh, but there were two curates. That was what made it so much the funnier." What about the charge that he may have killed all his wives? "He is really rather naughty sometimes," she says, laughing softly.

The two doctors are about to drag Smith away when Moon persuades them to leave him at Beacon House where they have formed their own High Court of Beacon that will try him. Smith seemingly tries to escape by taking over a hansom cab, but then he turns the cab around and returns.

The informal and illegal trial opens with Dr. Werner as the chief prosecutor. Moon and Inglewood act as attorneys for the defense.

The trial starts with the prosecution introducing two letters, each describing an incident that occurred when Smith was an undergraduate at a Cambridge college. The writers report seeing a philosophy professor, James Emerson Eames, sitting astride a flying buttress outside the professor's office balcony. On the balcony is Smith, holding a revolver. They see him shoot twice at the frightened philosopher.

The defense then produces a long document signed by both Eames and Smith. The professor, it reveals, was a great admirer of Schopenhauer, the eminent German pessimist who maintained that human life was so miserable that one is better off dead. Young Smith is in Eames's office listening to Eames argue that life is not worth living, a point of view Smith has found troubling.

Suddenly Smith produces a revolver, forces Eames to his balcony and then out on a Gothic gutter. He offers to put the philosopher out of his misery by shooting him. Eames, of course, is in no way anxious to die. Smith, who has won a prize at Cambridge for marksmanship, fires two bullets not far from Eames's head. When Eames begs for mercy, Smith lets him back in his office. The professor has now abandoned his pessimism. The document ends with the two signers saying they are adjourning to The Spotted Dog for beer.

Perhaps Chesterton got the idea for this scene from Boswell's life of Samuel Johnson:

> In 1769 Boswell mentioned to Johnson that David Hume once told him he was "no more uneasy to think that he should *not be* after this life, than he *had not been* before he began to exist."
>
> Johnson: "Sir, if he really thinks so, his perceptions are disturbed; he is mad. If he does not think so, he lies. He may tell you he holds his fingers in the flame of a candle without feeling pain; would you believe him?"
>
> Boswell: "Foote, Sir, told me that when he was very ill he was not afraid to die."
>
> Johnson: "It is not true, Sir. Hold a pistol to Foote's breast, or to Hume's breast, and threaten to kill him and you'll see how they behave."

In volume 14 of the Ignatius Press edition of Chesterton's *Collected Works*, editor Denis J. Conlon reprints some episodes G. K. had written in the late 1890s for a novel he called "The Man with Two Legs." The novel was never completed. One episode involves a man named Willis Hope who claims that one's existence is not a "benefit," and "I for one would certainly much rather do without—Good God, sir! Put that down!" The Reverend Eric Petersen is pointing a pistol at him.

"You wish to get rid of existence," the minster says, "allow me." Hope turns pale, cries "Help!" and obviously desires to go on living.

The Beacon High Court next considers the burglary charges against Smith. A letter is read from an Anglican clergyman Smith met at a meeting of Christian Socialists. The letter tells how he had accompanied Smith on a wild trek over rooftops during a dense London fog. They come to a house which Smith says he intends to "permeate." He opens a trapdoor on a roof and the pair descend into a living room lined with books.

Smith amazes the curate by locating some wine, and while they are enjoying the wine a young woman enters the room and Smith introduces her as his wife. It is, of course, his own house. Smith frequently breaks into his home like a burglar so he can see his house from a fresh perspective. There have been occasions when he was arrested for housebreaking.

Reading about this episode reminded me of a "burglary" Chesterton tells about in his autobiography. As a young man courting his wife to be, G. K. once retrieved her forgotten parasol from a deserted railway station. It was, he writes, his first and last burglary, and "very enjoyable."

We learn that Mrs. Smith has trouble keeping servants. Smith has a habit of knocking at his front door and asking a servant if a Mr. Smith lives there, and if so, what is he like?

The charge of deserting a wife is next considered by the Beacon House court. A gardener testifies that he once overheard Smith say to Mary:

> I won't stay here any longer. I've got another wife and much better children a long way from here. My other wife's got redder hair than yours, and my other garden's got a much finer situation; and I'm going off to them.

Smith then takes off down a road carrying a rake.

A letter from a French innkeeper is now read to the court. The writer owns a tavern by the sea. Smith steps out of a fishing boat, wades ashore, and tells the innkeeper he is searching for a house near a green lamppost and a red pillar-box.

Another letter, this one from Russia, is next read. It speaks of how Smith got off a train, told the writer he was looking for a certain house, but that it must be farther east. He leaves on the same train.

Wong-Hi, in a missive from China, tells how he was visited by a wild man holding a rake and looking for a house with a green lamppost and a red mail box.

Finally, a letter is introduced to the court from a housemaid who worked for Mrs. Smith. She says a giant suddenly appeared in the garden with a grimy rake—"a huge, horrible man, all hairy and ragged like Robinson Crusoe." All Mrs. Smith said to him was that he needed a shave. "The man looked around the garden and said 'Oh, what a lovely place you've got.'"

"He has stopped here ever since," the maid adds, "and does not give much trouble, though I sometimes fancy he is a little weak in his head."

Smith, the court now realizes, has circumnavigated the world so he could arrive back home and see his house, wife, and two children in a new light.

The court next takes up the charge of polygamy. Lady Bullingdon tells in a letter that a Polly Green once worked in her town as a dressmaker. She turns down a proposal by a local undertaker to marry a "village idiot" named Smith. The pair then leave town.

Several letters are next read about a Miss Blake, a stenographer. Smith, disguised as an organ grinder, likes to play outside her window. She likes to respond by rhythmic random typing. One day he carries her out of her office, puts her on his barrel organ, and wheels her out of town. She is never heard from.

Next comes a letter from the head of a girls' school. She learns that a man in the area named Smith is an authority on British names. She invites him to lecture on the topic. Everything goes well until Smith starts suggesting that every person with a place name should go live in that place. If a man has a trade name he should adopt that trade. If a surname is a color, one should wear that color. The girls participate in a wild discussion. What about persons with names like Low, Coward, and Craven? The girls raise problems of women with last names such as Mann and Younghusband.

Smith suddenly produces a hammer and some horseshoes, and declares his intention of starting a smithy in town. One of the young schoolgirls, a Miss Brown, is wearing a red-brown dress that exactly matches the color of her hair. Smith is so overwhelmed that he proposes on the spot. Miss Brown accepts, and a few days later she and Smith have vanished.

Michael Moon gives an eloquent final summation speech. All of Smith's so-called wives, he points out, had red hair. Moreover, all had last names that were colors. (Miss Blake was probably a mistake for a Miss Black.) He points to Mary Gray. There, he says, sit all of Smith's wives.

> This man's spiritual power has been precisely this, that he has distinguished between custom and creed. He has broken the conventions, but he has kept the commandments. It is as if a man were found gambling

wildly in a gambling hell, and you found that he only played for trouser buttons. It is as if you found a man making a clandestine appointment with a lady at a Covent Garden ball, and then you found it was his grandmother. Everything is ugly and discreditable, except the facts; everything is wrong about him, except that he has done no wrong.

It will then be asked, "Why does Innocent Smith continue far into his middle age a farcical existence, that exposes him to so many false charges?" To this I merely answer that he does it because he really is happy, because he really is hilarious, because he really is a man and alive. He is so young that climbing garden trees and playing silly practical jokes are still to him what they once were to us all. And if you ask me yet again why he alone among men should be fed with such inexhaustible follies, I have a very simple answer to that, though it is one that will not be approved.

There is but one answer, and I am sorry if you don't like it. If Innocent is happy, it is because he *is* innocent. If he can defy the conventions, it is just because he can keep the commandments. It is just because he does not want to kill but to excite to life that a pistol is still as exciting to him as it is to a schoolboy. It is just because he does not want to steal, because he does not covet his neighbour's goods, that he has captured the trick (oh, how we all long for it!), the trick of coveting his own goods. It is just because he does not want to commit adultery that he achieves the romance of sex; it is just because he loves one wife that he has a hundred honeymoons. If he had really murdered a man, if he had really deserted a woman, he would not be able to feel that a pistol or a love-letter was like a song—at least, not a comic song.

Everyone gathers outside the boarding house to celebrate the trial's end. Smith emerges turning cartwheels and shouting "Acquited! Acquited!"

A new gale is rising. Everyone is in a festive mood. Inglewood kisses Diana. G. K. is unable to resist the book's single note on his church's rhetoric. Catholic Mary says to Rosamund, "You go down the king's highway for God's truth, it is God's."

Smith and Mary now intend to visit an aunt who lives nearby to pick up their two children. Whenever they go on one of their "holidays," never longer than a fortnight, they leave the children with their aunt.

Manalive was first published in England in 1912 by Thomas Nelson, and in America the same year by John Lane. The Lane edition has a full color frontispiece by Dudley Tennent. Ignatius Press has an edition (it also includes *The Flying Inn*) with scores of enlightening footnotes by the editor. It is high time for Dover to reprint an inexpensive paperback.

Smith's trip around the world was anticipated in a story Chesterton wrote in 1896 titled "Homesick at Home." It tells how White Wynd (Wind) circles the globe to return to his home. Even the great wind that opens *Manalive* blows through White's hair as he starts his mad journey. The story seems to have been first published in volume 14 of the Ignatius Press edition of G. K.'s *Collected Works*.

Eleven reviews of *Manalive* are reprinted in D. J. Conlon's *G. K. Chesterton: The Critical Judgments* (1976), including one review that is eight pages long. Most of the reviews mix admiration with dislike. At one end of the spectrum is Rebecca West calling the novel an "intoxicating and delicious fairy tale." At the other end is an unsigned critic writing that the book is a "total, unredeemed, blank failure. It is the very worst of all of Mr. Chesterton's novels." It would be interesting to know if such divergent reactions (even today) to Chesterton reflect a division between those who believe in God and those who don't.

In *A Life in Letters* (letters by F. Scott Fitzgerald), edited by Matthew Brucoli (1994), Scott has high praise for Chesterton's fiction, including *Manalive*. In a letter to Edmund Wilson he speaks of his first novel, *This Side of Paradise*, as showing "traces of Chesterton."

NOTES

1. Children are grateful when Santa Claus puts in their stockings gifts of toys or sweets. Could I not be grateful to Santa Claus when he put in my stockings the gift of two miraculous legs? We thank people for birthday presents of cigars and slippers. Can I thank no one for the birthday present of birth?—Chesterton, *Orthodoxy*, chap. 4.

2. Ian T. Benson, in his fine introduction to *Manalive* (vol. 7 of the Ignatius Press edition of Chesterton's *Collected Works*), calls attention to G. K.'s essay "On Running After One's Hat." It can be found in Chesterton's *All Things Considered* (1955).

22.

THE NIGHT BEFORE CHRISTMAS

Burton Stevenson, in his book *Famous Single Poems,* considered the work of what he called "one-poem poets." These are poets who wrote many poems, all of which have been totally forgotten except for one poem, often to the puzzlement of the poet, which became famous. In many cases even the writer of such a poem is forgotten. How many persons, for example, can name the author of *Casey at the Bat*? I tell the story of this greatest of all baseball poems in my *Annotated Casey at the Bat,* a collection of parodies and sequels.

Langdon Smith, who wrote the ever popular poem *Evolution,* is an extreme case. It is his only known published poem! See my tribute to Smith in my book *Order and Surprise.*

Clement Clarke Moore is another of Stevenson's one-poem poets. He wrote a raft of other poems, of which he was proud, but not one of them ever appeared in an anthology. The following appreciation of his immortal ballad about Saint Nick was the introduction to my *Annotated Night Before Christmas,* a 1991 Simon & Schuster book recently reissued by Prometheus Books. The book is a collection of sequels and parodies, closing with a fine tribute by Chesterton to Father Christmas.

> Now don't laugh too loud. We have come through so many cynical years it gives me a chuckle to think what would happen to anyone in a New York night club who got up and recited my choice heart throb. . . . There are clinging to those illy-written lines a something that brings a real nostalgia of childjoy such as we will never know again. It is not maudlin; it is real. Am I wrong?
>
> —The artist James Montgomery Flagg, quoted in *Favorite Heart Throbs of Famous People* (Grosset and Dunlap, 1929), as he explains to Joe Mitchell Chapple, the book's editor, why he chose "The Night Before Christmas."

The story of how Clement Clarke Moore came to write his immortal poem, and the curious history of its early anonymous publications, have been told many times. Indeed, every Christmas they seem to be told again in some newspaper or periodical. No one, however, has told it better or in more detail than Burton Egbert Stevenson, an American novelist and anthologist, in his book *Famous Single Poems*.

Stevenson was fascinated by a common literary phenomenon that seems to interest nobody these days except me. "One swallow may not make a summer," he writes at the beginning of his introductory chapter, "but one poem makes a poet. Immortality may be—and often has been—won with a single song." Stevenson's term for rhymesters who achieve this is "one-poem men." In many cases such poets do their best to write "serious" poems in classical style, poems that can be skillfully crafted and are often gathered in books. All these strenuous efforts fade into oblivion while the single poem, to the author's amazement, strikes some sort of responsive chord in millions of readers that causes it to outlast by far everything else he or she wrote.

"It is not altogether astonishing that a masterpiece should live," writes Stevenson, "but, by some curious quirk, a mere jingle, which possesses no possible claim to inspiration, often proves more immortal than an epic. 'Bo-Beep' outlives 'Paradise Regained,' and grave and scholarly men, after a lifetime of labor in their chosen fields, have been astonished and chagrined to find that their sole claim to public remembrance rested upon a bit of careless rhyme written in a moment of relaxation." Stevenson continues:

> Poets have always been the special sport of Fortune, which delights to play with them, to whirl them aloft and to cast them down, to torment them with fleeting glimpses of happiness in the midst of long nightmares of despair, and especially to condemn their favorite children to swift oblivion and to raise up some despised and rejected outcast for the admiration of mankind. Nobody—poets least of all!—has yet discovered the formula which will assure immortality to a poem. Mere size will not do it—the most ambitious edifices are usually the first to crumble. Neither polished diction nor lofty thought will do it—most deathless songs are written in words of one syllable on the simplest of themes.

Another mark of a poem's popularity, one not mentioned by Stevenson, is that other versifiers are moved to produce sequels and imitations. In my *Annotated Casey at the Bat* I gathered as many such ballads as I could find about Mighty Casey, the hero of Mudville, who has become almost as permanent a fixture in American folklore as Santa Claus. Ernest L. Thayer, the author of *Casey*, had many other humorous poems published in the *San Francisco Examiner* and other Hearst newspapers, but only *Casey* survived. Stevenson devotes a chapter to this greatest of all baseball poems, and I relied heavily on it in the introduction to my *Casey* anthology, as I will rely here on his chapter about Moore's poem.

"The Night Before Christmas," as Moore's poem has come to be known, has far exceeded even Thayer's ballad in popularity and in the number of its sequels and parodies. No other poem by an American has been printed more often in newspapers, periodicals, and books, or has been illustrated by more graphic artists.

The illustrations run the gamut from such sophisticated artists as Arthur Rackham to the primitive simplicity of Grandma Moses, who did her paintings for the poem when she was one hundred years old. I do not know how many times the poem has been set to music, but I know of at least three examples. Sheet music for the ballad, composed by Hanna Van Vollenhoven, was published by the Boston Music Company in 1923. Sheet music for a Decca recording and arranged by Harry Simeone was published in 1945 by Shawnee Press, Delaware Water Gap, Pennsylvania. By far the most successful musical adaption was by the composer Johnny

Figure 19: Santa Claus by Thomas Nast.

Marks. The Saint Nicholas Music Company of New York City published the sheet music in 1952. Rosemary Clooney and Gene Autry recorded the song, as did Mitch Miller, the Ames Brothers, Gisele MacKenzie, and many others.

It is often claimed that Moore's poem has been translated into all the major languages. However, Anne Lyon Haight, in her preface to a catalog for an exhibition of her marvelous collection of early printings of Moore's ballad, has this to say:

> I know of no collection, large or small, of editions in other than English. I have advertised for foreign language copies, but with no success. I have written to scores of dealers across this land in an effort to obtain such editions. None was forthcoming. In my travels in Europe, and in other lands, I have sought fruitlessly for these elusive ghosts. The answer is always the same, even in Britain—no one has heard of the poem, nor of Clement C. Moore, and much less of any translation of it. I can only conclude that the reports of the existence of such translations—with one or two exceptions—are nothing less than myths, to be accounted as fabrics of fiction, as ghosts in our records of literature.

The two exceptions mentioned by Mrs. Haight are *Besuch vom Sankt Nikolaus*, a German translation that appeared in the December 21, 1949, issue of *Heute*, a magazine published by the Office of the US High Commissioner for Germany; and *Nuit de Noël*, printed in Paris the same year as one of Simon & Schuster's French series of Little Golden Books, with illustrations by Corinne Malvern. I am told that a later French translation, *La Veilee de Noël*, was published by Grosset and Dunlap in 1962, with pictures by a Japanese artist, Gyo Fujikawa.

Vincent Starrett, a Chicago writer, critic, and poet, was one of the first to amass a collection of book printings of Moore's poem. Many similar collections have been and are being made. Unfortunately, collectors of the poem are seldom interested in acquiring its imitations, and the task of running them down for this anthology has not been easy. There is no good way to search for them except in special collections. I have had to rely mostly on verse I happened to stumble upon and poems remembered by friends. I am under no illusion about the number of gems I have probably missed. Hundreds of parodies have no doubt been printed in newspapers, periodicals, and on Christmas cards since Moore's poem was first published.

Clement Clarke Moore was the only child of the Right Reverend Benjamin Moore, a bishop of the Protestant Episcopal Church in New York City and rector of Trinity Church on Wall Street. He was also president of Columbia College, now Columbia University. During the Revolution he never wavered in his loyalty to England. When Alexander Hamilton lay dying after his duel with Aaron Burr, it was Benjamin Moore who gave him the last rites. Benjamin's wife, Charity, inherited a large tract of farm and orchard land between Nineteenth and Twenty-fourth streets, extending from Eighth Avenue to the Hudson River, on Manhattan's West Side. It was in the family mansion on this property, within what is now called the Chelsea section of Manhattan, that Clement was born in 1779.

Moore received a bachelor of arts degree from Columbia in 1798. He had intended to become a minister but changed his mind, and devoted himself instead to classical and Oriental studies. His *Compendious Lexicon of the Hebrew Language*, two volumes published in 1809, was the first English-Hebrew lexicon printed in the United States. When the General Theological Seminary of the Episcopal Church was organized, he gave it the land between Ninth and Tenth avenues and between Twentieth and Twenty-first streets. Here the seminary buildings were built and still stand. Three years later he became a professor of Greek and Oriental literature at the seminary. After his death in 1863 he was buried in the Trinity Church cemetery at 155th Street and Amsterdam Avenue.

Every year, in late December, a Clement Clarke Moore Christmas Commemoration is held in the Church of the Intercession at Broadway and 155th Street in uptown Manhattan. After the candlelight service, at which Moore's ballad is read, there is a lantern procession, with luck through snow, to Moore's grave across the street. The 1991 Commemoration will be the eightieth.

It was in the winter of 1822 that Moore, in a lighthearted mood, dashed off his famous ballad to read at Christmas to his two daughters, seven-year-old Margaret and six-year-old Charity. (A third daughter, three-year-old Mary, was too young to appreciate the poem.) Present during the reading was either Harriet Butler, daughter of David Butler, then rector of St. Paul's Episcopal Church in Troy, New York, or a friend of hers. In either case, Harriet copied the poem in her "album," a book that young ladies of the time liked to keep, so that she could read it to the children at her husband's rectory.

Just before Christmas, a year later, an unknown woman (most likely Mrs. Butler) gave a copy of Moore's poem to the editor of the Troy *Sen-*

tinel, without telling him who wrote it. The poem was published on the second page of the Tuesday, December 23, 1823, issue with the title "Account of a Visit from St. Nicholas." It was preceded by the following note by Orville Luther Holley, the paper's editor:

> We know not to whom we are indebted for the following description of that unwearied patron of children—that homely, but delightful personification of parental kindness—Santa Claus, his costume and his equipage, as he goes about visiting the fire-sides of this happy land, laden with Christian bounties; but, from whomever it may have come, we give thanks for it. There is, to our apprehension, a spirit of cordial goodness in it, a playfulness of fancy, and a benevolent alacrity to enter into the feelings and promote the simple pleasures of children, which are altogether charming. We hope our little patrons, both lads and lassies, will accept it as proof of our unfeigned good will toward them—as a token of our warmest wish that they may have many a merry Christmas; that they may long retain their beautiful relish for those unbought, homebred joys, which derive their flavor from filial piety and fraternal love, and which they may be assured are the least alloyed that time can furnish them; and that they may never depart from that simplicity of character, which is their own fairest ornament, and for the sake of which they have been pronounced, by authority which none can gainsay, the types of such as shall inherit the kingdom of heaven.

The fifty-six-line poem was such an instant success with readers that for several years the *Sentinel* reprinted it each Christmas and about 1830 began issuing it as a broadside, with a woodcut by Myron King showing Santa sailing over rooftops in his sleigh. The sheet was handed out by carriers when they delivered the paper's Christmas edition.

Not until seven years later was the question raised about the poem's origin. Who was the author, the New York *Courier* wanted to know, when it printed the ballad on January 1, 1829. On January 20, Holley answered the query:

> A few days since, the editors of the New York *Courier*, at the request of a lady, inserted some lines descriptive of one of the visits of that good old Dutch Saint, St. Nicholas, and at the same time applied to our Albany neighbors for information as to the author. That information, we apprehend, the Albany editors cannot give. The lines were *first* pub-

> lished in this paper. They came to us from a manuscript in possession of a lady of this city. We have been given to understand that the author of them belongs, by birth and residence, to the city of New York, and that he is a gentleman of *more* merit as a scholar and a writer than many more of more noisy pretensions.

Holley's italicizing of *more* in his last sentence, and his use of it two more times, suggests that he knew who the author was, but respected Moore's desire not to have his name associated with what he considered insignificant doggerel. At any rate, the many reprintings of the poem over the next eight years, in all parts of the country, carried no byline. In 1837 *The New York Book of Poetry*, published by George Dearborn, contained the ballad and for the first time identified Moore as the author. Seven years later Moore included the poem in his own collection, a book titled simply *Poems*, published in New York by Bartlett and Welford. Most of the poems in this book are deadly serious. Here is how Stevenson describes them:

> He invokes the Muses, celebrates various nymphs, apostrophizes Hebe, Apollo, Terpsichore, Pallas and numerous other gods and goddesses, and capitalizes Fancy, Hope and so on, all in the good old way. He castigates the follies of the times, especially the freedom with which young ladies display their charms; decries the wine-bibber and exalts the drinker of water; writes at length (fifty pages) of a family excursion to Saratoga; tells of his sorrow at the death of his wife, and includes a few translations from classic poets.
>
> In a word, the volume is entirely characteristic of the times, when writing verses was a sort of courtly accomplishment with which the gravest men were supposed to amuse their leisure hours.

Some of the poems, notably the one about Saint Nicholas, were intended for amusement. In his preface, Moore apologized for them in the following letter to his children:

> My Dear Children:
>
> In compliance with your wishes, I here present you with a volume of verses, written by me at different periods of my life.

> I have not made a selection from among my verses of such as are of any particular cast; but have given you the melancholy and the lively; the serious, the sportive, and even the trifling; such as relate solely to our own domestic circle, and those of which the subjects take a wider range. Were I to offer you nothing but what is gay and lively, you well know that the deepest and keenest feelings of your father's heart would not be portrayed. If, on the other hand, nothing but what is serious or sad had been presented to your view, an equally imperfect character of his mind would have been exhibited. For you are all aware that he is far from following the school of Chesterfield with regard to harmless mirth and merriment; and that, in spite of all the cares and sorrows of this life, he thinks we are so constituted that a good honest hearty laugh, which conceals no malice, and is excited by nothing corrupt, however ungenteel it may be, is healthful to both body and mind. Another reason why the mere trifles in this volume have not been withheld is that such things have been often found by me to afford greater pleasure than what was by myself esteemed of more worth.

Not until 1848 was the poem published singly in an illustrated edition by the New York firm of Henry M. Onderdonk. Its woodcuts by T. C. Boyd are faithful to Moore's poem in showing reindeer the size of cats, and a Santa small enough to emerge from the huge chimneys and fireplaces of the time. Dodd, Mead reprinted this rare little book in 1971, adding a life of Moore, his wife, and relatives, and included a photograph of the poem as handwritten by Moore in 1862.

In 1897 William S. Pelletreau, in another small book titled *The Visit of St. Nicholas*, told for the first time the full story of how the poem was written and came to be published in the Troy newspaper. No one has questioned the authenticity of this account except the descendants of one Henry Livingston, born in 1748, who lived on an estate in Poughkeepsie, New York. The Livingston family stoutly maintained that their ancestor was the true author of the ballad. Somehow it found its way into Moore's hands, they claimed, and Moore was unable to disown it once it had been attributed to him. Stevenson carefully considers the Livingston claim and roundly rejects it. The sole basis for it seems to be that Livingston did write some verses in anapest meter, but, as Stevenson says, all anapestic verse sounds the same.

One of the most interesting things about Moore's classic is the extent to which it shaped America's greatest myth—almost its only great myth: that of Santa Claus.

True, Santa had his misty antecedents, but they bore little resemblance to him. Saint Nicholas, a fourth-century bishop of Myra (now Kale), a seaport in Asia Minor, was tall, thin, and dignified. Venerated throughout the middle ages, he even became a patron saint in Greece and Russia. His feast day on December 6 was a traditional holiday in Europe until the Reformation, when he fell into disrepute, especially in Protestant countries. Eventually his feast day merged with December 25, which since the fourth century had been Rome's traditional day for celebrating the birth of Jesus. The English word *Christmas* is a compression of the Roman Catholic "Mass of Christ."

Although Holland was Protestant, for some reason Saint Nicholas survived there as a Christmas gift-bringer. For six centuries Dutch children have put their shoes by the fireplace on St. Nicholas Eve (December 5), along with food for the saint's horse. During the night, Sinterklass and his Moorish assistant Zwarte Piet (Black Peter) arrive by ship from, of all places, Spain. The saint gallops through the heavens on his white horse, from roof to roof, with Black Peter somehow following. It is the Moor who pops down chimneys to leave gifts, while Sinterklass, not wanting to soil his white robe and red cassock, drops candy down the chimney and into the waiting shoes.

In other nations, other legendary gift-bearers make annual midwinter visits to homes. There is Father Christmas in England and France (where he is called Père Noël), though in recent decades Santa Claus and Father Christmas have become interchangeable names in England. Germany's history of Santa Claus counterparts is long and confusing. Shortly after the Reformation, Saint Nicholas was disguised as Christkindel (Christ child), and usually depicted with a halo and riding a white donkey. It is unclear whether this was the child Jesus or a boy or girl sent by Jesus. Accompanying the Christ child was a black-faced ogre known by various names, of which Knecht Ruprecht was apparently the most common.

According to Phillip Snyder's "St. Nicholas and His Counterparts," a chapter in his *December 25: The Joys of Christmas Past*, there were other legendary German gift-bringers, such as Pelznickel (Nicholas in furs)—the Pennsylvania Germans called him Belsnikel or Belsh Nickel—and Weihnachtsmann (Old Father Christmas). Early German settlers in Pennsylvania corrupted Christkindel to Kriss Kringle, apparently the most popular counterpart of Santa in Germany today. Snyder speaks of two early books published in Philadelphia: *Kriss Kringle's Book* (1842) and *Kriss Kringle's Christmas Tree* (1845). Today Kriss Kringle, both here and in Germany, is another name for Santa Claus.

In Italy, the gifts are distributed by the good witch Befana. According to an ancient Christian legend, Befana was sweeping her house when the Three Wise Men rode by and invited her to go with them to Bethlehem. Befana said she was too busy. Later, regretting her decision, she began wandering about the world under a curse that does not allow her to die. Each year on the eve of Twelfth Night (January 5), the day that commemorates the visit of the Magi, Befana slides down the chimney on her broom to fill shoes and stockings with candy and small toys, always peering into the faces of sleeping children, hoping to see the infant Jesus.

In Spain, on the eve of Twelfth Night, it is the Wise Men themselves, arriving by camels on their way to Bethlehem, who brings gifts to children.

In Russia, before the Communists took over, Befana had a counterpart—an old peasant woman called Babushka—and also Saint Nicholas, who would leave presents around a decorated fir tree on January 7 (in the Russian calendar). Communist leaders abolished the holiday. Now it is Grandfather Frost, accompanied by the Snow Maiden, who brings the toys on New Year's Eve. The tree is called a New Year's tree. With fresh winds starting to blow through the Soviet Union, perhaps Saint Nicholas and Babushka will make a comeback. In any case, Grandfather Frost is a fat old fellow in a red suit who looks just like Santa Claus.

There are no traditions of an annual gift-bringer in the Scandinavian countries. In the Orient, a small number of Christian groups have variants of Santa Claus that go by different names. Buddhist Japan, however, which has been eager for the past forty-five years to imitate American culture, has made Christmas a day of celebration—but not a work holiday. "Jingoru Beru" ("Jingle Bells") is played everywhere, and people wish one another a "Mari Kurisumasu." A few years ago I read in the *New York Times* that big department stores in Japan were dressing their elevator girl starters in miniskirted Santa Claus suits.

Papa Noël and Viejo Pascuero are, I am told, the Santa Clauses of Brazil and Chile, respectively. They bring gifts from the South Pole on

Christmas Eve, which in South America, of course, is the hottest time of the year. Canada, no doubt because of its proximity to the United States, has adopted Santa Claus entirely, except for French-speaking Quebec, where Père Noël is still the toy-bringer. In Australia, depending on where you live, it is either Santa Claus or Father Christmas who brings the gifts.

It was the Dutch settlers in New York, then called New Amsterdam, who brought Saint Nicholas to our shores, where his name was soon corrupted to Sint Klaes (also spelled Sinterklass, San Claas, and in other ways) and finally to Santa Claus. And it was Washington Irving who was the first to write about this. In his *History of New York* (1809), Irving reports that Dutch children would hang their stockings by the fire on St. Nicholas Eve (December 5), and the saint would come "riding over the tops of trees" in a "wagon" to send toys and candy rattling down the chimneys.

A Dutch ship sailed into New York Harbor with a wooden figurehead by Saint Nicholas—he was the patron saint of sailors—described by Irving as having "a low, broad-brimmed hat, a huge pair of Flemish trunk hose, and a pipe that reached to the end of the bowsprit." The Dutch settlers, he wrote, "built a fair and goodly chapel within the fort, which they consecrated to his [Saint Nicholas's] name; whereupon he immediately took the town of New Amsterdam under his peculiar patronage, and he has ever since been, and I devoutly hope will ever be, the tutelar saint of this excellent city." (All the above, by the way, turned out to be entirely fictitious.)

In book 1, chapter 5, of his history, Irving recounts a dream about Saint Nick that Moore must have seen since it describes smoke circling Santa's head and his gesture of putting a finger alongside his nose:

> And the sage Oloffe dreamed a dream—and lo, the good St. Nicholas came riding over the tops of the trees in that selfsame wagon wherein he brings his yearly presents to children; and he came and descended hard by where the heroes of Communipaw had made their late repast. And the shrewd Van Kortlandt knew him by his broad hat, his long pipe, and the resemblance which he bore to the figure on the bow of the Goede Vrouw. And he lit his pipe by the fire and he sat himself down and smoked; and as he smoked, the smoke from his pipe ascended into the air and spread like a cloud overhead. And the sage Oloffe bethought him, and he has-

> tened and climbed up to the top of one of the tallest trees, and saw that the smoke spread over a great extent of country—and as he considered it more attentively, he fancied that the great volume of smoke assumed a variety of marvelous forms, where in dim obscurity he saw shadowed out palaces and domes and lofty spires, all which lasted but a moment and then faded away until the whole rolled off and nothing but the green woods were left. And when St. Nicholas had smoked his pipe, he twisted it in his hatband, and laying his finger beside his nose gave the astonished Van Kortlandt a very significant look; then mounting his wagon he returned over the tree tops and disappeared.

Moore's poem was the second major influence on the evolving American Santa. Having Saint Nick come down the chimney in person was probably Moore's invention. Of course, only a small person could do that, and for this reason Moore's Santa is a "jolly old elf" who arrives on a sleigh pulled by "tiny reindeer." Early pictures of Santa show him small. Even as late as 1902, when William Wallace Denslow, the illustrator of L. Frank Baum's *The Wonderful Wizard of Oz*, illustrated Moore's poem, he drew Santa as a little elf. It was also Moore who gave Santa his twinkling eyes, rosy cheeks, nose like a cherry, little round belly, and a large pack of toys.

"The sleigh drawn by reindeer was pure invention!" exclaims Stevenson. This was widely believed until it was discovered that in 1821, a year before Moore wrote his masterpiece, a small hand-colored book of eight pages and eight pictures was published in New York by William B. Gilley, a friend and neighbor of Moore. It was titled *The Children's Friend: Number III. A New-Year's Present to the Little Ones from Five to Twelve*. No author's name is on the booklet, but it is now known to have been both written and illustrated by Arthur J. Stansbury, a Presbyterian minister. It is the earliest known Christmas book printed in the United States. One of its pages shows Santa in a sleigh pulled by a single reindeer. Underneath are these lines:

> Old Santeclaus with much delight
> His reindeer drives the frosty night
> O'er chimney tops, and track of snow
> To bring his yearly gifts to you.

It is hard to believe that Moore did not see this booklet before he introduced reindeer into his poem. Did the Reverend Arthur Stansbury invent the sleigh and reindeer? No one knows. It is possible that both sleigh and reindeer were part of the Dutch folklore about Saint Nick, though it was probably Moore who put the number of reindeer at eight.

Almost all fiction and verse published for children in England and America before 1800, with the exception of anonymous jingles like the Mother Goose rhymes, were didactic—intended to teach something, especially moral and religious values. I am indebted to Betsy Shirley for suggesting that not only was Moore's ballad the first American poem for children that has lasted; it also may have been the nation's first significant nondidactic poem for children.

The third great molder of Santa Claus was the cartoonist Thomas Nast (1840–1902), a German-born New Yorker who became famous for his attacks on the Tammany Tiger, a beast Nast created to symbolize New York City's corrupt political system presided over by Tammany Hall's "Boss" William Marcy Tweed. Nast also invented the donkey and elephant as symbols of our two major political parties. His first sketch of Santa appeared in *Harper's Weekly*, January 3, 1863, and from then until 1900 scarcely a Christmas went by without a Nast Santa in *Harper's Weekly*. Around 1869 Nast did full-color illustrations for an edition of Moore's poem published by McLoughlin Brothers.

Nast's earliest pictures of Santa show him small enough to go down chimneys, but in later Nast drawings he appears to be of normal height or even larger. He is always fat and jolly with a white beard. Nast replaced the all-fur coat Moore had given Santa with a red satin suit trimmed with white ermine. The pointed stocking cap, buckled shoes, and wide belt were other Nast touches. It was also Nast who gave Santa a home and workshop at the North Pole, and a large book in which he records the names of children who have behaved well throughout the year. Some of Nast's pictures show Santa answering phone calls from children and reading letters in which they request certain toys. The only Nast touch that did not survive was the sprig of mistletoe he always put on top of Santa's cap.

Hundreds of short stories and many novels have been written about Santa. Mrs. Claus seems to have made her first appearance in *Goody Santa Claus on a Sleigh Ride*, an 1899 novel by Katherine Lee Bates. Bates was a popular novelist in her day, but is now remembered only as the composer of "America the Beautiful." In my opinion, the best novel about Saint Nick is *The Life and Adventures of Santa Claus* by L. Frank Baum of Oz fame. Baum locates Santa in the Forest of Burzee, not far from Oz. The book explains how the gods of the forest reward Santa's goodness by conferring upon him the Mantle of Immortality.

It has become customary these days to complain about how early American Christmases, and the happy Christmases of Charles Dickens, have become corrupted by tasteless contemporary greed. For a growing number of families, the days preceding Christmas are now days of anxiety and depression. Yuletide Blues, as they have been called, are responsible for a rise in suicides that seems to peak during the period between December 25 and January 1. In big cities, the Christmas Blues are especially harsh on those who live alone, and on the thousands of homeless.

Things are not much better among well-to-do families. Like Thanksgiving, Christmas has degenerated into a holiday on which we gorge ourselves with fattening food, unmindful of the millions dying of starvation throughout the world. Traffic deaths in the nation increase by the hundreds as more people take to the roads, often under the influence of alcohol or drugs. Wild office parties have become such drunken saturnalias that many firms ban them.

Weeks before Thanksgiving, department stores now launch their seasonal orgy of hawking gifts that nobody needs and may not even want. Each year parents feel obligated to spend more and more on toys for children whose demands are intensified by television commercials that are often highly deceptive. To add to the annoyance, large toys often require complicated assembling according to opaque instructions, using screws and bolts that are sometimes missing or don't fit. A week later, most children have grown tired of their new possessions.

"Forgive us our Christmases," wrote Carolyn Wells, whose humorous verse included relentless attacks on Yuletide commercialism, "as we forgive those who Christmas against us."

Christmas lasted only a few days in Dickens's time, Russell Baker pointed out in a 1976 *New York Times* column, but "nowadays it persists like an onset of shingles. You spend a month getting ready for it and two weeks getting over it. . . . If Scrooge . . . had started dreaming on November 25 and spent the next four weeks being subjected to desperate sales clerks and electronically amplified 'Jingle Bells,' he probably would have stopped at the Cratchits' on that fateful evening only long enough to smash Tiny Tim's little crutch."

Commercialism has so overwhelmed the story of the birth of Jesus that some fundamentalist churches today oppose the celebration of Christmas. Early Calvinists in Europe and England and our own pre-Revolutionary Puritans were similarly offended by Yuletide shenanigans. Salem, Massachusetts, actually passed a law in 1628 banning the celebration of Christmas as a "wanton Bacchanalian feast. . . . God's time must not be frittered away." And today, three of our largest fundamentalist sects—the Seventh-Day Adventists, Jehovah's Witnesses, and the Worldwide Church of God—openly resist any celebration of it. For some fundamentalists, Christian Christmases are the work of Satan himself. In 1989 my son Tom was handed a tract issued by a nondenominational firm in Bennett, North Carolina, that bitterly denounced Santa Claus. Santa is shown on the tract's front page in a red suit, with horns on his head, a red tail, and a pitchfork in hand. Inside, we are told that "by the early eighteen hundreds Christmas had become so idolatrous that the American poet Clement C. Moore decided to take Christ completely out; by putting Santa in his sacrilegious poem, 'The Night Before Christmas,' and thus placing Santa on the Christmas throne and Christ in the cradle." Accompanying these sentiments is an anonymous five-stanza poem titled "Ho! Ho! Ho!" It begins:

> The devil has a demon,
> His name is Santa Claus.
> He's a dirty old demon.
> Because of last year's flaws.
> He promised jack a yo-yo,
> And Jill a diamond ring.
> They woke up Christmas morning
> Without a single thing.

And ends:

> One day they'll stand before God
> Without their bag of tricks.

> Without their red-nose reindeers,
> Or their phony Old Saint Nicks;
> For Revelation twenty-one,
> Verse eight, tells where they'll go;
> Condemned to everlasting hell,
> Where there'll be no Ho! Ho! Ho!

I'm surprised that the author of this tract failed to observe that the letters of SANTA can be rearranged to spell SATAN!

It would be interesting to know how many parents today tell their small children that the Santa Claus in whose lap they sit for a photograph is the real thing, and how many explain that he is just someone dressed like Santa. The poor child must surely be confused if he or she is taken to more than one shopping mall, each with its own Santa, and then sees other Santas on sidewalks clanging bells for the Salvation Army. This leads to a question about which I have no firm opinion. Is it good or bad to let children believe in Santa Claus?

There are persuasive arguments on both sides. Parents who think it is bad maintain that it is never good to tell children lies. When children learn the truth about Santa, they will find it harder to believe other things their parents tell them. A religious parent can argue that, after finding out Santa isn't real, a child will naturally conclude when he is older that God, too, isn't real—only a mythological figure to go alongside such cultural icons as Saint Nick and Uncle Sam.

On the other hand, it can be argued that children adore fantasy and derive enormous pleasure from the Santa Claus myth. True, they soon learn that the old fellow is a fraud, but this is harmless disenchantment. Moreover, children who are told there is no Santa come in conflict with believing pals if they try to reveal to them the awful truth. If you are a secular humanist (a popular euphemism for atheist) you can argue that letting children swallow the myth for a brief time is good training for becoming adult skeptics about God and Jesus. After all, the great biblical miracles strain credulity even more than the story of a fat man who comes down chimneys and enters millions of houses in a single night.

British-born Robert Service (1875–1958), in *Rhymes of a Rolling Stone*, has a short poignant poem titled "The Skeptic" that goes like this:

My Father Christmas passed away
When I was barely seven.
At twenty-one, alack-a-day,
I lost my hope of heaven.

Yet not in either lies the curse:
The hell of it's because
I don't know which loss hurt the worse—
My God or Santa Claus.

Gamaliel Bradford's essay "Santa Claus: A Psychograph" is a spirited defense of keeping the Santa myth alive among children. His final paragraph is worth quoting:

> So the legend of Saint Nicholas is a lovely and delectable myth, the last living relic of the vanishing world of dreams. The fairies are gone. No little children or innocent maidens watch any longer through the ardent summer nights to catch some echo of the songs and dances of tiny people, footing it daintily over the dewy turf. The witches are gone. Unpleasant old ladies can look about them ill-favoredly and purvey gossip without the danger of being burned at the stake. Nobody pays heed to them and nobody fears what they do. The ghosts are gone. Solitary graveyards are far more comfortable and agreeable sojourning places in the summer evenings than crowded streets where one has to be constantly on the watch against becoming a ghost oneself. Santa Claus alone still lingers with us. For Heaven's sake, let us keep him as long as we can. There are some excellent people who are scrupulous about deceiving their children with such legendary nonsense. They are mistaken. The children learn to see soon enough, too clearly and too well, or to think they do. Ah, leave them at least one thrill of passionate mystery that may linger with them when the years begin to grow too plain and dull and bare. After all, in this universe of ignorance, anything may be true, even our dreams.
>
> And there is a still deeper value in the preservation of the Santa Claus legend, even by those who have no faith in that or any other legend whatever. For such preservation typifies the profound principle that, sacred as both are, the law of love is higher than the law of truth. For this there is a perfectly simple and unassailable reason, that truth at

its best is deceiving, but love is never. We toil and tire ourselves and sacrifice our lives for the dim goddess Truth. Then she eludes us, slips away from us, mocks at us. But love grows firmer and surer and more prevailing as the years pass by.

Therefore, why should not old and young alike, in the brilliant, deceptive Christmas moonlight, hearken for the tinkling bells and the pawing reindeer and echo the merry greeting of the saint, broadcast to the whole wide world:

"Merry Christmas to all and to all a good night."

I happen to be a philosophical theist, so let me toss out a suggestion surely made before, though I have not encountered it. "Great believers," Thornton Wilder liked to say, "are great doubters." It's a poor faith that can't preserve itself in the face of evidence which seems to point toward foolishness. Perhaps allowing children to believe in Santa Claus, then later telling them that Santa doesn't exist, is a healthy preparation for adult trust in a power higher than imaginary gods and devils. A faith that can be damaged by early disenchantment over Santa Claus surely is not much of a faith.

23.

THE GREAT CRUMPLED PAPER HOAX

I have two ineradicable prejudices. I dislike all modern poetry that lacks music. Such music is traditionally supplied by rhyme and/or metre, and devices such as onomatopoeia, alliteration, and other techniques. But almost all contemporary English and American poets are incapable of adding music to their work. Their poems consist of prose, often obscure, divided into lines to make it look like poetry. Where today is there a Homer, a Dante, a Milton, a Keats, a Byron? If great poets exist, their work has never found a publisher. Lord Dunsany put it well in a lecture on poetry. Poems, he said, should ring like bells, but all modern poetry does is go klunk.

My other prejudice is against a subset of abstract art that goes by the name of minimal. I'm thinking of the all-black or all blue canvasses of Ad Reinhardt, or Carl Andre's pile of bricks. My spoof on minimal sculpture, "The Great Crumpled Paper Hoax,"appeared in the Summer 2007 issue of the *Ojai Orange*.

"The trouble with your art," said Hazel, my significant other, "is that you don't have a gimmick."

It was a hot day in July. I was sitting opposite Hazel in a small basement bar in Soho, a district on Manhattan's lower east side. Above us was the Archibald Gallery where fourteen of my paintings were hanging. It was my first one-man show.

The exhibit had been a colossal failure. Only one newspaper, the *New York Times*, covered the show. The art critic who reviewed it called it "the most vapid show" he had seen in decades. Not a single picture sold.

"It's not just that you don't have a gimmick," Hazel went on. She had an annoying habit of always saying exactly what she was thinking even when it pained a listener. "It's not just that you need something to distinguish your work from everybody else. It's that you never learned how to draw."

I put down my glass of beer and winced. Hazel was telling it like it was. I couldn't paint a decent-looking cow if my life depended on it.

"You're right as usual, my love," I said. "But what can I do? I refuse to get up at six every morning to go to a job I hate—a job that can't lead anywhere."

"If you want my advice," she said, "which I doubt, forget about landscapes and realism. Go abstract. If your painting is totally nonobjective nobody can tell whether it's good or bad. But you *have* to have a gimmick. You have to have something everyone will recognize as your trademark."

"Gimmick's a good word for it," I said. "Did you steal it from that magician friend of yours? For years I've racked my brain for a fresh gimmick. But it ain't easy to think of one. I can't drip paint like Pollock, or draw two rectangles like Rothko and paint each a different color. I can't slash the canvas with fat black brush strokes like Kline, or paint the canvas a solid color like Reinhardt, or glue broken dishes to the canvas, or decorate it with elephant dung."

Hazel flourished her empty glass as a signal to the bartender. "Well, keep trying. Have you thought of moving from paint to minimal sculpture?"

"Like Andre's pile of bricks?"

"Yes, like Andre's pile of bricks."

I paid the tab and we parted, Hazel to her basement brownstone apartment, I to my lonesome loft in Brooklyn.

Next morning, during breakfast, I read another review of my show. It was even more scathing that the first. The *Brooklyn Eagle* wondered—was the show a deliberate joke?

I removed the double pages, crumpled them into a ball, and hurled the ball across the room. It struck a wall, then bounced to the uncarpeted floor where a sunbeam from a skylight turned the air to dust and the paper to gold.

"It's beautiful," I said to myself. "It sparkles like a large yellow diamond." I was reminded of G. K. Chesterton's quip: "all is gold that glitters."

Suddenly a bizarre thought bombarded my brain. I leaped to my feet. "Eureka!" I shouted. "I've found my gimmick!"

With trembling hands I picked up from the kitchen counter a vertical spike, mounted on a wood base, that I used for spearing grocery receipts. I removed the receipts, retrieved my crumpled sphere, then pushed it partway onto the spike.

Next, I sprayed the ball a bright blue. From a distance it looked like a NASA photograph of the earth from outer space, only an earth with a lovely crumpled surface.

The following day I showed the thing to my cousin Archibald. "Magnificent!" he exclaimed. "My gallery is booked through August and September, but in October I can display, say, a dozen of your works. Of course you'll have to change your name."

I changed my name from Joseph Johnson to Francis Feemster. The October show was a huge success. The *New York Times* critic warbled about how my crumpled paper modeled the crumpled state of the earth's environment. All twelve of my spheres were sold. MOMA bought a huge ball that I made from a complete Sunday edition of the *Times*, using glue to fasten the outer layers. For the Brooklyn Museum of Art I provided a large blue ball mounted on a spike, in turn mounted on a wooden cube that contained a tiny motor and two AA batteries. The motor rotated the ball slowly from east to west.

The following year I shifted to my pink period. This was followed by a multicolored period using crumpled Sunday comic pages. A Chicago manufacturer bought rights to mass produce the balls in colored plastic. Of course they sold for a price much lower than one of my originals. *Time* devoted three colorful pages to what they called crumpled paper sculpture. Feemster became famous. After Hazel and I were married, we moved to a high-rise apartment on Charles Street in the Village. A well-known art critic is taping conversations for a biography.

I'm now in my black period. The black symbolizes the black future of old earth.

It goes without saying that Hazel and I have been extremely careful never to let on that crumpled ball art is a put-on. The deception continues to distress us. Like Pollock, I'm drinking more booze than I should, and my dear wife is hinting that maybe it's time for me to check into a detox facility. If I drink myself to death, she tells me, she'll see that a crumpled ball of concrete will rest on top of my tombstone.

24.

SO LONG OLD GIRL

In the fall of 1945, when the USS *Pope* (DE 134) and its four surviving sister ships tied up in Florida's Green Cove Springs harbor for decommissioning, I wrote this poem. I am not much of a poet, but for sentimental reasons I decided to keep this piece from dropping into oblivion. It would be easy to touch it up, but I have left it exactly as it came out of the ship's typewriter except for substituting "blacks" for "negroes."

Before the war, when I worked in the University of Chicago's press relations office, I had known about the efforts by Enrico Fermi and his associates to achieve a nuclear reaction that would make an atom bomb possible. It was deep night in the North Atlantic, while I was standing a lookout's watch on the Bridge, that news came over the intercom about the bombing of Hiroshima. I knew at once that the war with Japan was over, and I was the only man on the ship with the slightest inkling of what had happened. It is this intimation about the future of warfare that gives my farewell to the ship a bit of nontrivial interest.

Four years ago
At the shipyard of the Consolidated Steel Corporation
In Orange, Texas,
Under the glare of blue arclights

Crackling from the welding tools of husky young Texas girls,
Slowly the steel ribs arose.
Of 180 men in the commissioning crew
Only 16 were "regular" Navy.
Most of them had never been on a ship before—
Just a random collection of farm hands, truck drivers, clerks, and short-order cooks,
With very little enthusiasm
For getting salt in their ears.

Proudly down the Sabine River to the Gulf
She sailed,
And at the River's mouth, on the Fourth of July, 1942,
Rolled her first depth charge—
For fun.
All hands crowded the weather deck to hear the explosion
And feel the deck shudder against their feet,
And watch the water break into a shimmering pattern
As the first shock waves reached the surface,
Then burst into a tall spout of water.

It was all part of the Big Game.

But it wasn't long until the cans were dropped
Not to make funny splashes
But with the express purpose of blasting holes
Into the sides of German submarines.

Remember how you shoved your thumbs hard
Against the red and green buttons
That fired the cans from the port and starboard K guns?
And later, when the Germans who survived
Were fished out of the spermy water
Leaving swirls of crimson on the oily surface,
You reflected that two dozen men had been suddenly strangled
By the pressure of your two thumbs.

It was only a year ago that a sister ship
Built at the same time in Orange,
And serving as part of the same killer group,
Blew in half in the North Atlantic when two fish
Caught her amidship.
There was no time to set her charges on "safe,"
So most of the men who survived the torpedoes
Were killed in the water by their own depth charges
When the ship sank to the proper depth.

But it wasn't enemy action that came closest
To destroying the ship.
It was a wave.
The storm—off the coast of Iceland—lasted four days,
And the wave was over fifty feet high.
It ripped a searchlight from the bridge,
Smashed the port lookout station,
And tore off gun shields like cardboard.
Members of the crew found bibles in the bottoms of their lockers,
Shook off the cockroaches,
And tried to find appropriate sections to read.
One seaman had to be held
While a pharmacist's mate gave him a shot in the arm
To quiet him down,
And on later cruises, when the sea was rough,
Old hands would walk around with grins
And say to the new men,
"You shoulda been on here when . . ."

And remember the German prisoners we picked up
From a surrendered U-boat?
And how they patted our depth charges
With foolish smiles on their white faces?
And how anxious we were to stand watch over them
So we could shoot them if they tried anything,
Then how confused we were later

When we discovered they were just a handful of ordinary guys
Like us?
But they thought Hitler was a 4–0 Joe,
And were surprised to see we had five Jewish sailors
And four blacks on the ship
Who ate and slept with the rest of us.
(They would be interested to know that last week
When we threw a final ship's party at the Mayflower Hotel in Jacksonville,
The four blacks couldn't attend
Because it violated hotel rules.)

And remember the night you sat along on the fo'c's'le,
Your feet resting on the anchor chain;
A full moon throwing a path of silver on the quiet sea
And the ship dark except for the pale violet glow
Of a vacuum tube visible through the porthole of the pilot house?
Dolphins playing around the bow, their wet backs
Shining in the milky light,
Made soft whistling noises
As they cut graceful arcs through the air,
And the ship herself cut the water
With a smooth, undulating, sensual motion.
For the first time the vessel lost her gray ugliness
And seemed more than just a hunk of steel
Welded together for war purposes,
And you understood for the first time
Why ships are spoken of in the feminine gender.

But that was long ago.
The ammunition has since been removed,
Proper preservatives applied to the guns,
Compartments sprayed with paint, dehydrated, and sealed,
And you can spin the wheel in the pilot house
Like a wheel at a carnival.
She is ready for the last paper to be signed,

And you have your orders to the Jacksonville Separation Center,
And you wonder why the Navy bothered to preserve the 3-inch guns
Because they are only little pop-guns now,
And the ship is about as much use in the next war
As a paper ship in a bathtub.
(But it is not prudent for an enlisted man
To question the ways of the Navy.)

So Goodbye old girl, sleep peacefully on the St. Johns River,
And if they ever tow you out and sink you
In an atomic bomb test,
Give my regards to your sister
And to the bones of the men who rest with her.
They'll be happy to know
That when the atom bombs start to fall
They are in a spot that is fairly safe.

PART THREE: L. FRANK BAUM

25.

QUEEN ZIXI OF IX

I first learned to read by looking over my mother's shoulder while she read aloud *The Wizard of Oz.* L. Frank Baum's fourteen Oz books were the great delights of my childhood. I not only owned all fourteen, I also acquired all of Baum's fantasies about magic lands outside Oz. Like so many children who became "Ozmapolitans," I drew a map of Oz before the Wogglebug's map was published. Decades later I became a founder of the International Wizard of Oz Club, the history of which is sketched here in chapter 31.

The magazine *Fantasy and Science Fiction* published my "Royal Historian of Oz," the first of many biographies to come. It is reprinted in *Order and Surprise,* followed by two articles on why so many librarians of the time had a low opinion of Baum.

It is hard now to believe, but in 1957, when Michigan State University reprinted *The Wizard of Oz,* the head of Detroit's public libraries explained why no city library carried an Oz book. They were crudely written, he said, unfit for today's children. A Detroit newspaper was so incensed by his remarks that it serialized *The Wizard,* each episode preceded by a paragraph saying that this is the book no Detroit child can find in a city library.

Librarians no longer bash the Oz books. Not after the Judy Garland movie, and a raft of scholarly books about Baum and Oz, such as Michael Hearn's massive classic *The Annotated Wizard of Oz* (Norton, 2000). The *Baum Bugle,* the official organ of the Oz club, is now a handsome periodical. The critics of children's literature have finally discovered, what of course the children always knew, that Baum was the Lewis Carroll of American letters. It is no coincidence that the first word of the first *Alice* book is Alice, and the first word of the first Oz book is Dorothy.

I provided introductions to seven Dover paperback reprints of Baum fantasies, several of which are reprinted here. Many critics consider *Queen Zixi of Ix* to be Baum's finest non-Oz fantasy. My introduction was for the Dover edition of 1971.

In the fall of 1905, when *Queen Zixi of Ix* was first published by the Century Company, New York, Lyman Frank Baum was well on his way to becoming this country's leading author of fantasy for children. He had published his first two Oz books, *The Wonderful Wizard of Oz* and *The Marvelous Land of Oz* (both available as Dover reprints), and a variety of other books including five non-Oz fantasies: *A New Wonderland* (also a Dover reprint, under its later title *The Magical Monarch of Mo*), *Dot and Tot of Merryland*, *The Master Key*, *The Life and Adventures of Santa Claus*, and *The Enchanted Island of Yew*. Baum had also written two books of short stories: *Mother Goose in Prose* and *American Fairy Tales*. All these books, and others of less importance, had been published within a seven-year period.

Queen Zixi of Ix was first serialized in the children's magazine *St. Nicholas*, from November 1904 through October 1905, with illustrations by Frederick Richardson. A Chicago-born artist, Richardson had been with the *Chicago Daily News* for fifteen years before moving to New York City in 1903. The pictures he drew for *St. Nicholas* were used again in the book, except for one picture showing Zixi begging the fairy queen to grant her wish. *Queen Zixi* was the only book of Baum's illustrated by Richardson, although he did provide the art for one of Baum's short stories, "A Kidnapped Santa Claus" (*Delineator*, December 1904).

St. Nicholas was owned by the Century Company. In a 1912 letter to Reilly & Britton, Baum's Chicago publishers, Baum speaks of having sold the serial rights of *Queen Zixi* for $1,500. This included the Century Company's right to publish it in book form. The 1905 first edition, of which two states have been distinguished, had a light green cloth cover with pictures on the front and back in red and dark green. It was first reprinted by the Century Company in 1906 and that same year a similar edition was issued in London by Hodder & Stoughton.

The book is dedicated to Frank Joslyn Baum, the oldest of Baum's four sons. Russell P. MacFall, in his biography of Baum, *To Please a Child* (Reilly & Lee, 1961), which he wrote with the aid of Frank, quotes Baum's inscription in the copy of *Queen Zixi* that he gave to his son: "You are the last of my boys to have one of my books dedicated to you, but it was your own wish, and in waiting for this story perhaps you have not been unwise. In some ways *Queen Zixi* is my best effort, and nearer to the 'old-fashioned' fairy tale than anything I have yet accomplished."

Queen Zixi is indeed the most classical in form of all Baum's book-length fairy tales, and in MacFall's opinion, Baum's best book next to *The Wonderful Wizard of Oz*. Edward Wagenknecht, in his 1929 booklet *Utopia Americana* (the first critical study of Baum's Oz books), as well as in later writings, has praised *Queen Zixi* as one of the best fairy tales ever written by anyone. It is in direct line with earlier fantasies in which palace life is combined with fairy lore and outrageous humor. One thinks of Andrew Lang's *Prince Prigio* (1889) and its sequel *Prince Ricardo* (1893), which Baum may have read and consciously imitated. (*Prince Prigio* was in turn a deliberate imitation of Thackeray's *The Rose and the Ring*; Lang even mentions that his mythical land of Pantouflia is near to Thackeray's kingdom of Paflagonia.) Some of the traditional fairy tale elements that Baum adopts in *Queen Zixi* are noted by MacFall: the magic wishing device used foolishly, the moralizing against vanity, the cruel foster mother (Aunt Rivette), and the Cinderella child (Fluff).

The name of Baum's kingdom, Noland (no-land), reminds one of Samuel Butler's *Erewhon* (with two letters switched, "Erewhon" is "nowhere" backward) and the Neverland (in the play, Never-Never Land) of J. M. Barrie's *Peter and Wendy*. According to the official map of Oz and its environs, by artist Dick Martin and cartographer James Haff, Noland is just across the northeast corner of the Deadly Desert that surrounds the rectangular-shaped Oz. Noland is bounded on the southeast by Merryland, on the west by Ix. (After Oz and Mo, Ix was Baum's third attempt to catch the public's fancy with a two-letter word of his own invention; he was still reluctant to continue his Oz series.)

Ix and Noland are separated by high mountains, though not as high as the giant stair-step mountains in the northern part of Noland where the evil Roly-Rogues lived before they bounced down to capture Nole, the capital of Noland. The river into which these green, yellow, red, and brown soccerball monsters were finally rolled remains nameless, but the map by Haff and Martin shows that it carried them into the Nonestic Ocean which Dorothy was destined to cross in the third Oz book, *Ozma of Oz* (1907). They finally settled on Roly-Rogues Island, not far from Noland's northern coast.

Figure 20: All the Roly-Rogues were thus rolled into the river, where they bobbed up and down in the water.

The Forest of Burzee, which Baum had previously introduced in his *Life and Adventures of Santa Claus*, is across the Deadly Desert opposite the central southern boundary of Glinda's Quadling region of Oz. Glinda and Ozma later attend a conference there, with the fairies of Burzee, in Jack Snow's *The Magical Mimics in Oz*.

Queen Zixi, Bud, and Fluff are guests at Ozma's birthday party at the close of Baum's fifth Oz book, *The Road to Oz*. Dorothy guesses Zixi's age to be sixteen, but the Wizard whispers that she is actually thousands of years old. (The Wizard was either exaggerating or mistaken; we are told on page 103 of the Dover edition that Zixi's age is a mere 683.) In Baum's *The Magic of Oz*, Kiki Aru, in the form of a hawk, flies over Noland and Ix, with brief stopovers in each before he continues his westward flight to Ev. Noland and Ix are also mentioned in several post-Baum Oz books. Boxwood, an Ixian forest ruled by Chillywalla and inhabited by creatures made of boxes, is visited by the adventurers in Ruth Plumly Thompson's *The Silver Princess in Oz*. In her earlier book *The Wishing Horse of Oz*, Miss Thompson tells of Skampavia, a small country bordering on both Ix and Merryland. It is ruled by King Skamperoo, a power-hungry scamp who tries to conquer Oz after finding himself unwilling to invade nearby Ix because Zixi is much too friendly and pretty. Although the Roly-Rogues do not reappear in any later Oz book, their ability to roll and rebound was used again by Baum in his next non-Oz fantasy, *John Dough and the Cherub*; one of its creatures is Para Bruin, a rubber bear who likes to bunch himself into a ball and bounce from high places.

The wordplay in *Queen Zixi* is typical of Baum. The king's five fat counselors have names that differ only by the five vowels. On page 20 we are told that if the bell is tolled the people will be told of the king's death. Tollydob, the lord high general, becomes on page 91 a lord *very* high general when he grows ten feet tall. Zixi's assumed name, "Miss Trust," implies that she is to be mistrusted, as Baum makes clear on page 186. Bud's exclamation, "Fudge!" (page 210), is another amusing bit of wordplay. There is a touch of Carrollian logical nonsense on page 97. After Ruffles, a shaggy dog, has reminded Tallydab that talking dogs are common in fairy tales, he is told by Tallydab that "this isn't a fairy tale, Ruffles. It's real life in the kingdom of Noland." The old law of Noland that provides for giving the crown randomly to the forty-seventh person who enter the city's gate is splendid whimsy, but is it really any more random or whimsical than the ancient practice, once believed by millions to be divinely ordained, of passing the crown to a king's eldest son?

Richardson, who illustrated *Queen Zixi*, was a slightly built, gray-eyed man who studied art at the St. Louis School of Fine Arts in St. Louis, Missouri, and the Academie Julien in Paris. For seven years he taught at the Chicago Art Institute. He was strongly influenced by the art nouveau movement that flourished in the 1890's in England, Europe, and America, and which has had a recent revival in the United States in the interest in Aubrey Beardsley, Tiffany glass, and the swirling, writhing patterns of psychedelic poster art. Selections from Richardson's elaborate pen-and-ink sketches for the *Chicago Daily News* were published in Chicago by the Lakeside Press in 1899 with the title *Book of Drawings*.

Queen Zixi was probably the first children's book Richardson illustrated. Later he did many others. The most elaborate was his *Mother Goose*, published by Volland in 1915 and still obtainable under an M. A. Donohue imprint. It contains more than a hundred full-color plates. Richardson also illustrated an *Aesop's Fables*, a *Pinocchio*, a volume of tales by Hans Christian Andersen, a little book of familiar short stories (such as "The Three Bears") called *Old, Old Tales Retold*, a book of Indian folk tales, a book of Japanese folk tales, and many other books for children including a series of elementary school textbooks called *The Winston Readers*. He illustrated Frank Stockton's *The Queen's Museum and Other Fanciful Tales*. Richardson died in New York City on January 15, 1937, survived by Allan Barbour Richardson, one of his two sons, now retired and living in Winchester, Virginia.

Baum's single attempt at science fiction, *The Master Key*, weaves an adventure story around several marvelous scientific inventions which are given to a boy by the Demon of Electricity and misused by him. The story's moral, even more appropriate today, is that science is rapidly providing "gifts" too dangerous for humanity, in its present immature state, to handle wisely. I doubt if Baum had a similar moral in mind for the magic cloak in *Queen Zixi*, but it is amusing nonetheless to take the cloak as a symbol of science. The golden thread that gives the shimmering cloth its magic power is the method of induction. The fairies of Burzee offer the cloak to suffering mortals to make them happy; instead, it is used so foolishly that the immortals are obliged to withdraw it.

Pressing this metaphor, we can see General Tollydob's great height as the awesome power science has given today's generals. Aunt Rivette's wings are symbols of airplanes and spacecraft. Jikki's six ubiquitous, identical servants are the mass-produced robots of an automated age; they are even numbered like the names of modern computers. Tellydeb's long arm is, of course, the long reach of science, not only into the depths of space but also into the depths of the microworld. Tillydib's royal purse, always filled with gold, represents the fabulous riches of a scientific technology. Even the talking ability of Ruffles—to twist the metaphor still more—calls to mind the efforts of certain zoologists to converse with dolphins. Of course, the allegory breaks down at many points. The cloak's magic power, for example, fails to work if the cloak is stolen, whereas the secrets of science obviously work whether stolen or not.

The main moral of *Queen Zixi*, and one that Baum clearly intended, is that it is vain and foolish to desire the impossible. The moral is underlined by Zixi's conversations with the owl who wants to be a fish, the alligator who weeps crocodile tears because he cannot climb a tree, and the ferryman's little girl who longs to be a man. Zixi is, of course, afflicted with the same foolishness. The golden-haired, black-eyed girl is not content to remain young and beautiful in her external appearance; she wants also to stay forever young and beautiful in a more fundamental sense than is belied by her haglike reflection in a mirror. She is the symbol of that ultimate of vanities, the desire of a mortal to escape earthly mortality.

In 1914 Baum's own motion picture firm, the Oz Film Manufacturing Company, made and released *The Magic Cloak of Oz*, a five-reeler based on *Queen Zixi*. Fluff was played by Mildred Harris, a twelve-year-old girl who four years later became Charlie Chaplin's first wife. The film—surprisingly good considering its time and how quickly it was made—closely follows Baum's text and Richardson's illustrations.

The book's first sentence, with "of Oz" added to "fairies," opens the screen play. Departures from the book include a beast called the Zoop, brief glimpses of the wooden Woozy (from *The Patchwork Girl of Oz*, which Baum had filmed earlier), and a band of robbers who steal Nikodemus, Aunt Rivette's donkey. (The evil robbers are first seen playing jacks, with a caption telling us that they had previously stolen "a nice little girl named Mary.") A variety of technical tricks liven the film, such as running it backward to show the defeated Roly-Rogues rolling back *up* their mountain instead of into a river.

I wish to thank James Haff, David Greene, Dick Martin, Fred Meyer, and Allan Barbour Richardson for generously offering information and advice in preparing this introduction.

26.

THE ENCHANTED ISLAND OF YEW

This appreciation of one of Baum's non-Oz fantasies originally ran in the *Baum Bugle*, Spring 1990.

In 1903, when *The Enchanted Island of Yew* was first published, L. Frank Baum could not have anticipated that his fame would ultimately rest on the series of Oz books that would follow *The Wonderful Wizard of Oz*. Apparently he saw his first Oz book as a fantasy complete in itself, demanding no sequel. *A New Wonderland* had been published in 1900 (the same year as *The Wizard*) and told about the magic country of Phunnyland—a name he later changed to Mo. *Dot and Tot of Merryland* (1901) introduced still another region of fantasy, and that year also saw the appearance of *American Fairy Tales* and Baum's one science fiction novel, *The Master Key*. In 1902, his *Life and Adventures of Santa Claus* added the Forest of Burzee and the Laughing Valley to Baum's growing list of imaginary places. It was not until 1904 that he returned to Oz with *The Marvelous Land of Oz*. Even then he did not think it important to reintroduce Dorothy, who presumably was permanently back in Kansas.

Yew (did Baum have in mind the word "you"?) can be found on the International Wizard of Oz Club's official map of countries near Oz as a pie-shaped island in the Nonestic Ocean, a few miles off the east coast of Oz. Patrick Maund has conjectured that Baum may have intended "Isle of Yew" as a play on "I love you." Its quartering into four kingdoms, with a fifth in the center, resembles the geographic divisions of Oz. Dawna, the eastern kingdom of Yew, obviously derives its name from dawn. Auriel, the name of the western kingdom, suggests sunset colors and Ariel, the airy sprite in Shakespeare's *Tempest*. Plenta, the name of the southern country, surely refers to its plenitude of fruits and flowers. What Baum had in mind, if anything, when he called the north country Heg is not clear. Spor, the mountainous central kingdom, suggests Sporades, the ancient Greek name for a group of islands in the Aegean, or the word "sport," but these may be no more than coincidences.

We know that Baum loved to indulge in wordplay when he invented proper nouns. King Terribus and Prince Marvel are obvious. Wul-Takim, king of the thieves, surely is a pun on "will take 'em," and Kwytoffle, the humbug sorcerer, is just as clearly a pun on "quite awful." Does Lord Nerle have any subtle reference? And was Baum aware of the French "merde" when he gave the name "merd" to Lady Seseley's father?

Like most of Baum's fantasy novels, *The Enchanted Island of Yew* is a sequence of exciting, amusing adventures, in this case based on the familiar plot line of the good knight and his faithful companion who ride randomly about to battle evil. (One thinks of Don Quixote and Sancho Panza, of the Lone Ranger and Tonto, and of a hundred other fictional pairs.) Unlike the old tales of knight errantry, Baum adheres to his usual practice of avoiding bloodcurdling, nightmare-producing violence by presenting his scoundrels as comic characters in the manner of the Nome King of the Oz series.

In almost all cases, the knaves on the enchanted island turn into admirable fellows. Wul-Takim becomes an honest man. King Terribus, whose evil nature sprang from shame over his ugliness, becomes a benevolent ruler after Prince Marvel rearranges his facial features. The giant Red Rogue of Dawna, after a lucky escape from the magic mirror, becomes an honest gardener. Kwytoffle, whose dictatorial control of Auriel rested on his threats to turn enemies into grasshoppers or June-bugs, is exposed as a harmless fake, and banished after eighteen lashes, one for each letter in "grasshopper" and "June-bug." All ends happily in the kingdoms visited by Marvel and Nerle. Nerle's masochism, which provides a running joke throughout the story, is finally cured after he comes to realize that pain is not as delightful as he had supposed.

The royal Dragon of Spor, whose ears flap like blankets on a clothesline, is one of the funniest of Baum's several comic dragons. He can't breathe fire because the wind has blown out his flame. Ordered to destroy Marvel and Nerle, he refuses on the grounds that his father (who got himself killed by Saint George) had raised him to be a gentleman. Nerle relights the dragon's breath with a match. This sudden appearance of a box of matches in what Baum had earlier described as the "old days," when there were no trains, phones, furnace smoke, books, or "inventions of any sort to keep people keyed up," is one of Baum's inspired comic touches. The dragon leaves "without undue confusion" to take his cough medicine. The scene is one of high humor, rivaled only by a later episode in which Kwytoffle, the fat little humbug wizard, keeps inventing lame excuses for being unable to turn his visitors into grasshoppers and June-bugs.

Passages of Carrollian nonsense pepper the narrative: after being told he can't sing while being hanged because the rope will cut off his breath, a thief remarks he will whistle instead. Baum admired the *Alice* books. Spots here and there suggest their influence, such as the two infernal mirrors that imprison viewers, and Prince Marvel's remark, echoing the Cheshire Cat, that "it really doesn't matter which way we go, so long as we get away from the Kingdom of Spor."

We have already mentioned how Yew's geography resembles that of Oz, and of course Kwytoffle resembles the Wizard of Oz before he learned real magic from Glinda. There are other similarities and even explicit references to Baum's earlier books. The unnamed tiny fairy, after her adventures as a knight, finds it good to return home like Dorothy, content with her memories. In chapter 6, Nerle recalls having met a "cowardly tiger." In the first chapter, we learn that although mortals cannot become fairies, "there was once a mortal who was made immortal"—a reference to Baum's *Life and Adventures of Santa Claus*.

It is worth observing that the plot of *Yew* is a sort of inversion of the plot of *The Wizard*. Dorothy is a mortal living a dull life in the gray, bleak plains of Kansas. Temporarily, she enjoys wild adventures among immortals. In this book, the situation is reversed. An immortal, who finds her life boring, enjoys exciting adventures for a year among mortals. Baum had little interest in Christian theology, so there are no intimations of the Incarnation, but there are parallels with ancient legends about the adventures of gods who descend to earth disguised as humans. The theme of the fairy girl living for a time as a mortal boy is a theme Baum would repeat in his second Oz book, where the immortal Ozma lives for a while as the boy Tip.

Scattered through *Yew*, as through all of Baum's fantasies, are asides that raise ethical and philosophical issues. "It is what we are not accustomed to that seems to us remarkable," says the double Ki. In context, the remark conveys the difficulty of one culture in understanding an alien culture, one of Baum's favorite themes. "How little one can tell from appearances what sort of heart beats in a person's body!" exclaims Prince Marvel after discovering that the fierce-looking Ki are really friendly, and the gentle-appearing Ki-Ki are not.

Baum's description of Twi is an astonishing tour de force. If you think carefully, you can find all sorts of logical contradictions in a land where everything is double, like seeing the world through special glasses that produce double images. Somehow Baum manages to pull it off. Note how cleverly he avoids a sky with a single sun and a single moon by bathing Twi in perpetual twi-light. This is one of several puns that relate to doubling. The Twis consider their visitors extremely "singular." Nerle wants to leave the place "double quick." Dancers entertain the visitors with a "double shuffle."

The Twi chapters introduce deep philosophical questions about personal identity. Baum had done this before in his account of how the Tin Woodman's body parts were slowly replaced by tin. Is he really the same person he was before? In *The Tin Woodman of Oz* Baum sharpens the paradox by having the tin man converse with his former head. Who is the real person, the head or the tin man? To make things worse, the witch who enchanted the woodman's axe also enchanted the sword of a Munchkin soldier. It kept cutting off parts of his body until he, too, was made entirely of tin. Parts of the bodies of the two men were glued together to make a composite "meat person" who married the former sweetheart of the woodman and soldier!

In Twi, the mystery of personal identity is focused on the double Hi-Ki, the high "kings" of Twi. They are two beautiful girls who think, talk, and behave as a single entity. When Prince Marvel separates them the results are calamitous. They become passionate enemies who declare war on each other. Peace is not restored until the prince removes the spell and they become a single mind again. All this may be amusing to a child, but it leads into profound metaphysical difficulties, some of which I explore in my *Whys of a Philosophical Scrivener* (Morrow, 1983).

Michael Hearn has suggested that Baum carefully contrived the episodes of *Yew* with a musical in mind that might repeat the enormous success in 1902 of the stage version of *The Wizard of Oz*. The scenes in Twi would be marvelous on stage, as well as the dramatic sex changes of Prince Marvel, played of course by an attractive young woman in tights. Such plans may have been shelved later in 1903 when Edith Ogden Harrison, wife of Chicago's mayor Carter Harrison, asked him to write a musical extravaganza based on her book *Prince Silverwings and Other Fairy Tales* (1902).

The musical was never produced, but Baum did complete the book. Although he borrowed some of the characters and scenes from Mrs. Harrison's tales, the plot was all his, with some of its ideas taken from *Yew*. (For a history of this ill-fated play and its characters and plot, see "The Faltering Flight of Prince Silverwings," by Michael Patrick Hearn, David L. Greene, and Peter E. Hanff, in the *Baum Bugle*, Autumn 1984). The three-act, eight-scene play features a little girl named Mayre, from the American town of Centerville, and her pet cat Tommy Bakfentz (back fence). Mayre is kidnapped by the immortals and, throughout the play, longs like Dorothy to go home.

The play opens in a fairy bower where Queen Charminia is angry with one of her messengers, Prince Silverwings. She removes his wings and orders him to perform an extraordinary act of kindness before he can be restored to her favor. The name "Kwytoffle" is given to the Gnome King (not to be confused with the Nome King of Baum's later Oz books) who runs an underground factory that supplies volcanoes with their fire and smoke. Kwytoffle, by the way, is an associate of Kwytkool, the Ice King and father of Jack Frost.

Baum reintroduced Marvel in "The Fairy Prince," a short play written entirely in rhymed couplets for children to present in a toy theater. The "playlet" was published in a juvenile magazine called *Entertaining* (December 1909). It was to be the first of a series of playlets by Baum, but the magazine folded before the series could continue. Set in the Forest of Burzee, the play opens with Princess Marvel weeping in the bower. She is approached by two mortal girls, Bessie Bodkin and Ruth Rutledge:

> Do not weep, my pretty fay;
> What has grieved you? Tell us, pray!

Because the princess has been "saucy" to "Queen Lulea, her punishment is to become a mortal until she performs a "noble deed." The two girls transform her into a boy who carries a sword. When Jack Turpin, a highwayman, tries to rob the girls, he is slain by Prince Marvel. The good deed now done, Marvel's original form is restored and the Queen welcomes her home.*

If you care to know something about Fanny Y. Cory, who illustrated this book and Baum's earlier *Master Key*, you will find an excellent article about her by Douglas G. Greene in the Spring 1973 issue of the *Baum Bugle*. She was one of the earliest and most successful of women artists who illustrated books for children. Born in Waukegan, Illinois, in 1877, she grew up in Helena, Montana, and attended art schools in Manhattan. She began drawing for *St. Nicholas* and, although this magazine published more of her work than any other, she also contributed to many adult magazines and illustrated numerous children's books. In 1903, she left her home in New Jersey to return to Helena, where she married Fred W. Cooney, the son of a wealthy rancher.

To pay for the education of their three children during the Great Depression, Fanny Cory began drawing a syndicated cartoon about the misadventures of a small boy named Sonny. *Sonnysayings*, as the strip was called, was first distributed by the Philadelphia *Ledger*. King Features took it over in 1935, and that same year she began drawing the syndicate's *Little Miss Muffet* strip. She continued both features until she retired in 1956 to Camino Island, Washington, where she lived until her death in 1972 at the age of ninety-four.

It was no accident that the first cover of *Yew* is so sexually alluring. In her youth, Miss Cory was a striking beauty with a splendid figure and a great zest for life. When she drew for *St. Nicholas*, she told Greene, she found the magazine over pious. "I was sometimes too brazen for the art editor, I think. I'll never forget the time they made me lengthen the toga I put on a little Roman boy in an illustration for a story. My version was too immodest." The *Miss Muffe*t strip, written by someone else, seemed to her much too dull to interest children. "She must be a relief to

---

*The playlet appeared in *L. Frank Baum's Juvenile Speaker* (Reilly & Britton Co., 1910; in 1911 the title was changed to *Baum's Own Book for Children*). The book also contained episodes from *The Enchanted Island of Yew*. The playlet was reprinted in the *Baum Bugle* (Christmas 1967).

mothers," Cory complained to an interviewer, "but sometimes I think she's too pure."

In 1936, Fanny Cory wrote and illustrated her own book for children, *Little Me*. Greene quotes one of its verses:

If God wouldn't keep his eye on me,
Day out and day in,
There's a lot of fun I'd have
Living in sin.

Great Kika-Koo! What strange sentiments to find in a child's book! Perhaps Fanny was recalling what Prince Marvel said to Lord Nerle at the close of chapter 5: "If we mistrusted all who have ever done an evil act, there would be fewer honest people in the world. And if it were as interesting to do a good act as an evil one, there is no doubt every one would choose the good."

I wish to thank Michael Patrick Hearn, author of *The Annotated Wizard of Oz* and currently working on an authorized biography of Baum, for lending me his copy of *Prince Silverwings* and for informing me that a play version of *The Enchanted Island of Yew* was presented on July 12, 1987, at the White Barn Theatre, in Westport, Connecticut.* The book was adapted for the stage by dancer-actress Carmen de Lavallade, who also played all the parts. Geoffrey Holder, her husband, directed and designed the play. Holder is the Trinidad actor and painter who had earlier directed and designed the Broadway production of *The Wiz*. Incidental music for *Yew* was provided by the Holders' son, Leo.

"I was poking around a bookshop," Carmen said in an interview published in the *Advocate and Greenwich Times* (July 12, 1987), "and I came across this old book. I guess I noticed it first because it had a pretty little illustration of a fairy on the cover. . . . As I started reading it, I realized that it was just charming. Usually Frank Baum is Oz, Oz, Oz, but this was different. . . . Like a lot of fairy tales, it is a story that appeals to children but that is really for adults. There are wonderful characters, and it is very funny. It is all about magic and the magic in nature."

*An earlier dramatization by Mary Isabel Buchanan was published by Samuel French in 1937.

POSTSCRIPT

After this introduction was published in the *Baum Bugle* (Spring 1990), Ruth Berman sent me a copy of a syndicated newspaper article titled "Romantic Marriage of the Girl Who Draws Cute Babies: Won Like the Heroine in a Melodrama." It appeared in Ruth's hometown paper, the *Minneapolis Tribune*, May 1, 1904. The article tells of a December occasion on which Miss Fanny Cory was at a skating party on the ice of Lake Sewell, near her home in Helena. The ice broke, she dropped into the freezing water, and was rescued by Fred Cooney. Two months later they announced their engagement.

27.

JOHN DOUGH AND THE CHERUB

This reprints my introduction to a Dover paperback edition (1974) of Baum's novel. Although not about Oz, it is one of Baum's best and funniest fantasies.

After Lyman Frank Baum's great success in 1900 with *The Wonderful World of Oz*, and his equally astonishing success two years later with the stage musical based on the book, Baum was at the height of his fame and creative energy. His second Oz book, *The Marvelous Land of Oz*, in many ways even better than its predecessor, was published in 1904 by Reilly and Britton, a small Chicago house that would publish all of Baum's Oz books and almost all of his many juveniles not about Oz.

Edward William Bok (later famous for his autobiography the *Americanization of Edward Bok*) was then editor-in-chief of *Ladies' Home Journal*. We know from a 1912 letter of Baum to his publisher, Frank Reilly, that sometime before 1906 Bok met with Baum and offered him $2,500 for serial rights to a new fantasy.[1] Baum responded with an early draft of *John Dough* in which Chick the Cherub did not appear. Bok returned the manuscript, asking Baum to add a human child to the tale.

"I had either a grouch or the big head," Baum said in his letter to Reilly, "and refused to alter the text." But after some second thoughts he decided that Bok was right. He rewrote the story, introducing a child with whom young readers could identify, and gave the manuscript to Reilly.

The book's working title was *John Dough, the Baker's Man.* John Rea Neill, who had illustrated the second Oz book (and would illustrate all subsequent Oz books by Baum and his successor, Ruth Plumly Thompson, as well as three Oz books of his own), did the pictures for *John Dough.*

In a copy of the first edition of *John Dough and the Cherub*, published by Reilly and Britton in 1906,[2] Baum wrote the following inscription to his son Robert: "Too bad this wasn't an Oz book, but I like the story just as well. This was the *first creation* of a gingerbread man and John Dough was original with this story."

Why did Baum emphasize "first creation"? Dick Martin, in his article cited in the footnote, gives the reason. The words imply that he invented John Dough before the production in New York in 1906, the very year Baum's book was published, of a musical comedy called *The Gingerbread Man.* The play and lyrics were by Frederick Ranken, the music by A. Baldwin Sloane. "It, too," writes Martin, "involved a gingerbread man magically brought to life, and (to twist the arm of coincidence a little further) *he* was also named John Dough. Here the parallel ends—the plot and characters of the Ranken-Sloane musical are quite different from those of Baum's book. On the other hand, the 1902 musical comedy of *The Wizard of Oz* bore little resemblance to Baum's original book—so perhaps, there is a connection—and a mystery yet to be solved."

John Dough and the Cherub is not, in my opinion, among Baum's best fantasies, but that doesn't mean it is not worth reading. It is typical Baum, funny and exciting, packed with Ozzy characters and episodes, and with outrageous surprises on almost every page. The book has its spots of humdrum writing, and some of its ideas are hackneyed, but it is hard to imagine a young reader, even today, who would be bored by the tale.

There are many reasons for supposing that Baum hoped to make a musical out of the story. He had tried without success to put his second Oz book on the stage (as *The Woggle-Bug*, which had a short "woggle," as one reviewer put it, in 1905) and in 1913 he would try again with his musical, *The Tik-Tok Man of Oz*. Michael Hearn, author of *The Annotated Wizard of Oz*, believes that many of *John Dough*'s defects can be attributed to Baum's intent to dramatize it. The episode on Pirate Island, for instance, does nothing for the plot, but pirates had been a big success in

Maud Adams's *Peter Pan* and would have added considerable color to a stage version. Of course any hope for dramatizing *John Dough* was dashed by the Ranken-Sloane production.

Ironically, it is Chick, added to the story as an afterthought, who dominates the narrative. Is Chick a boy or a girl? Baum does not tell us. All masculine and feminine pronouns are avoided (often awkwardly, as when Chick is referred to as "it"), and at the book's close, when Chick grows up as the Head Booleywag (Prime Minister) of Hiland and Loland, we still do not know if the Booleywag is a man or a woman.

Baum and his publisher exploited the mystery for all the publicity they could get.

Figure 21: Announcement of the publication of *John Dough* placed in major magazines.

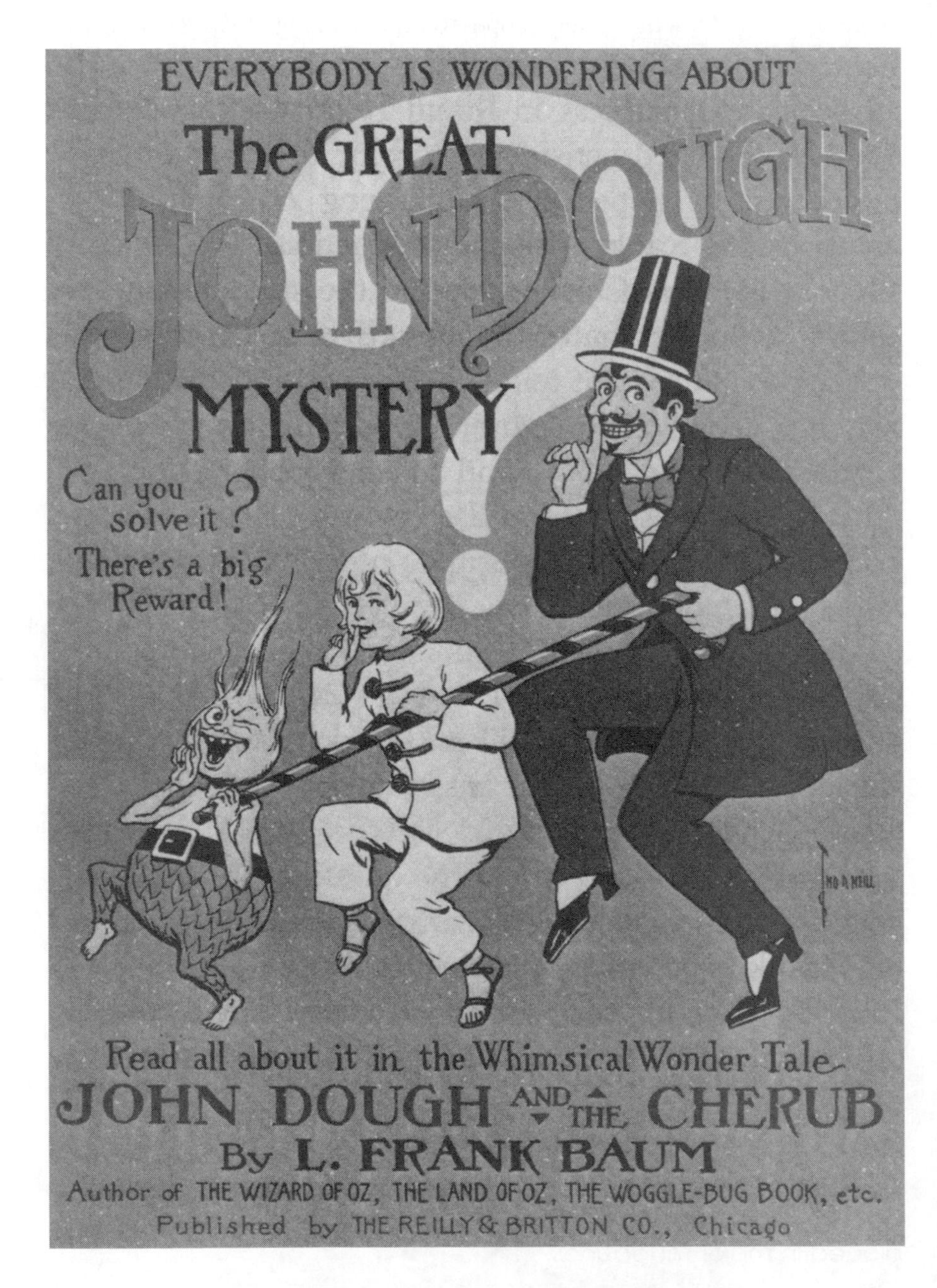

Figure 22: Book poster by John R. Neill for *John Dough.*

A mustard-colored contest blank, tipped in the book's early printings, offered cash prizes for the best statement of why readers thought Chick a boy or a girl. It is amusing to learn that the contest left the question unresolved.

Read This Before You Read the Book

The Great

JOHN DOUGH MYSTERY

Is the Cherub Girl or Boy?

$500.00 for the Best Answers

The Owner of this Book if not Over 16 Years Old is Eligible to Compete

The publishers of **John Dough and the Cherub** are in doubt as to whether the Cherub is a girl or a boy. The author, Mr. L. Frank Baum, may know, but if so, he has not told us. We have, therefore, determined to allow Mr. Baum's little friends to decide the question, and offer $500.00, divided into 135 gifts, for the best solutions of the mystery.

Chick, the Cherub, is one of the two most important personages in the story and the character is complex and many-sided. Some of Chick's traits seem to indicate that he is a boy, while others point to her being a girl. Some of the expressions Chick uses lean one way and some another. But read for yourself and send in your answers and reasons.

HOW THE GIFTS WILL BE AWARDED

A committee consisting of Mr. Baum, Mr. Henry M. Hyde, author and editor, and Mr. Wilbur D. Nesbit, of the Chicago Evening Post, will select from the answers received the 135 which give the **best reasons** for the conclusion that Chick is a boy or is a girl; the majority of these will determine the answer as to the first question. The gifts will be awarded to the 135 children in the order of the excellence of the reasons given. Awards will be made and the gifts distributed on or before January 15th, 1907, as follows:

LIST OF GIFTS

1 gift of		$100.00
1 " "		50.00
2 " "	$25.00	50.00
9 " "	10.00	90.00
22 " "	5.00	110.00
100 " "	1.00	100.00
		$500.00

To Question No. 1, the answer must be but one word—**Boy** or **Girl**.

To Question No. 2, the answer must **not** contain more than 25 words, but as many less as desired.

Only children not over 16 years old may compete. The blank below **must** be used in competing. Fill in the blank and mail before January 1st, 1907, to

THE REILLY & BRITTON CO., PUBLISHERS, CHICAGO

NOTE—The above offer expires December 31st, 1906. Answers not written on blanks from books, and not mailed before January 1st, 1907, as shown by postmark on envelope, will not be considered.

Tear off the blank here

THESE ARE THE QUESTIONS

1. Is Chick a Boy or a Girl? ANSWER__________
2. Why do you think so? ANSWER (in not over 25 words)__________

CANCELED

My name is __________ I live in __________

At (street address)__________

State of __________ and am _____ years of age.

Mail this slip before December 31st, 1906, with your answers plainly written in ink, to

THE REILLY & BRITTON CO., DEXTER BUILDING, CHICAGO

Figure 23: Sample of slip inserted in early printings of *John Dough*.

Although no documentation has yet been found, Baum's son Frank, who collaborated with Russell P. MacFall on a biography of his father (*To Please a Child*), told MacFall that the first prize of $100 was divided between two contestants. One asserted that Chick was a boy, the other that Chick was a girl.[3]

Several press clippings about the contest are preserved in Baum's scrapbooks. One newspaper story (see the *Baum Bugle*, Spring 1967) compares the mystery to Frank Stockton's famous "Lady or the Tiger?" story, then continues with what surely is a fabricated conversation. Asked by his publisher if Chick is a boy or girl, Baum reacts with amazement. "Doesn't it tell in the story?" Informed that it does not, Baum replies, "I cannot remember that Chick the Cherub impressed me as other than a joyous, sweet, venturesome and loveable child. Who cares whether it is a boy or a girl?"

Unsatisfied, the publisher questions his office staff only to get contradictory opinions. A second appeal is made to Baum. All he will say is, "Leave it to the children."

Another clipping, reproduced in the same issue of the *Baum Bugle*, shows three pictures of Chick, drawn by Neill to promote the story's serialization in 1906 newspapers. In the center picture, Chick wears the sexless pajamas and sandals that Chick wears in the book. On one side we see how the child would look if dressed like a boy, on the other, if dressed like a girl.

Is the Cherub a Girl or a Boy?

Figure 24: Promotional material drawn by John R. Neill for the 1906 newspaper serialization of *John Dough*.

John Dough provides little information about Chick's background aside from the fact that Chick is the world's first incubator baby. This explains Chick's residence on the Isle of Phreex (Freaks) and probably why Baum chose the name "Chick" (as Baum well knew—his first book was on chicken rearing—incubators were used for hatching chicken eggs long before they were used for prematurely born human babies.) Like most of Baum's child protagonists, nothing is said about Chick's father or mother. Indeed, the implication is that the Cherub *has* no parents. We do know that Chick is at least eight, blonde, blue-eyed and curly-headed. The child is always happy, always frank, clever, "wise for one so young," creative, brave, unprejudiced, and friendly. Chick likes to use the latest slang. Anyone the child particularly admires is "all right." One of the Cherub's talents is producing an ear-splitting whistle.

Above all, Chick is an adventurer of the open road. The child doesn't care where it is (page 286). Chick is equally unconcerned with what happens: "I'm not afraid. Anything suits me" (page 275). "What's the use of staying outside, when the door's open?" (page 298). "It doesn't matter where we go, so long as we keep going" (page 273). Compare that last remark with this dialog from Jack Kerouac's *On the Road*: "'Where we going, man?' 'I don't know but we gotta go.'"

In sum, Chick is a sandal-footed highway freak—a flower child of the counterculture, self-sufficient, androgynous, parentless, and happily "into" oatmeal and cream instead of drugs.

Chick's companion, John Dough (an obvious pun on John Doe), is the book's principal "non-meat" personage. He is a life-size gingerbread French gentleman, with top hat and candy cane, made by Jules Grogrande (*gros* and *grande*?), a Parisian baker who has settled in an unidentified American city. The powerful Arabian elixir which brings John to life makes him wise and strong, and capable of speaking all languages, modern and classic. (Since John Doe is everyman, Hearn has observed, it is natural that he speaks all languages.) He suffers occasionally from soggy feet, chipping, and loose glass eyes. Like the Scarecrow and Tin Woodman, he neither eats nor sleeps, though he *is* capable of drinking. He is deemed "all right" by Chick because, in spite of his dread of being eaten, he restores Princess Jacquelin to health by allowing her to nibble the stump of his left hand. (Note how cleverly Neill conceals John's mutilated hand in most of the pictures, after John's fingers have been eaten by Ooboo, and before the hand is finally restored.)

Para Bruin, the third member of the book's unlikely trio of adventurers, is one of Baum's most lovable creations. He is made of indestructible para rubber, hollow like a rubber ball, kind and harmless (how could his rubber teeth harm anyone?), and a ham vaudevillian who loves to roll up like a ball and bounce from high places to amuse the crowd. It is to his credit that he doesn't love everybody. I can still remember the satisfaction I felt as a child when Para Bruin bounced down from the sky to demolish Sport, surely one of the most unpleasant characters in all of Baum's fantasies.

Figure 25: Illustration by John R. Neill for *The Road to Oz*, showing King Dough, the Head Booleywag, and Para Bruin.

Chick, John Dough, and Para Bruin attend Ozma's birthday party at the close of *The Road to Oz*. His "Gracious and Most Edible Majesty" brings Ozma a gingerbread crown as a gift. When Dorothy asks Button-Bright if Chick is a boy or girl, Button-Bright responds with his usual, "Don't know." Para Bruin is amazed by all the strange people he sees in Oz. Button-Bright asks if John Dough is good to eat.

"Too good to eat," says Chick, and the Scarecrow assures John that visitors to Oz are never eaten. Chick informs Billina, a yellow hen, that it (Chick) never had any parents.

"My chicks have a parent," the hen replies, "and I'm it."

"I'm glad of that," says Chick, "because they'll have more fun worrying you than if they were brought up by an incubator. The incubator never worries, you know."

There are scores of lesser characters in *John Dough*, both meat and meatless. Duo, the two-headed dog, anticipates the Pushmi-Pullyu of Hugh Lofting's *Dr. Dolittle* books.[4] Sir Austed Alfrin is a spoonerism on Sir Alfred Austin, poet laureate of England in 1906. Tietjamus Toips, whose symphony is harder to understand than one by Vogner (Wagner), plays on the name of Paul Tietjens, a friend of Baum who composed the music for *The Wizard of Oz* musical. The name is also a pun on "pajama tops." Is Sir Pryse Bocks, inventor of the rain-repelling tube, a spoonerism on "Bok's price"—Bok's demand that Baum add a child to the tale? More likely (as David Greene called to my attention) the name refers to the popular prize contest which the *Ladies' Home Journal* sponsored every month. Suggestions were mailed to "Mr. Bok's personal box," and (Greene adds), with "Sir" before the name it becomes "surprise box." Maria Simpson, the name of the Lady Executioner, is so artificially presented that Green thinks it must refer either to someone readers of 1906 would recognize or to one of Baum's personal friends.

On page 233 John Dough, angry at the macaw who is laughing at him, calls the bird a "rampsy." What is a rampsy? The answer lies in an obscure spot: a short story called "Nelebel's Fairyland" which Baum wrote exclusively for the *Russ* (June 1905), a college paper published in San Diego. The story (reprinted in the *Baum Bugle*, Christmas 1962) gives added facts about the immortals who live in the Forest of Burzee, south of Oz, and who figure prominently in Baum's two earlier fantasies, *The Life and Adventures of Santa Claus* and *Queen Zixi of Ix*. It seems that Queen Lulea, "annoyed at the awkwardness of the huge gigans, transformed them into rampsies—the smallest of all immortals." So far as I know, it is the only other reference to rampsies in all of Baum's writings.

There are four fat ladies in *John Dough*: Madame Tina, the baker's wife; Bebe Celeste, one of the freaks of Phreex; the mother of the Princess; and the Lolander who bakes a new hand for John Dough the

First. Baum is no doubt reminding his readers about the hazards of eating too much French pastry.

Taking cues from some references to Hiland and Loland in the *Magic of Oz*, James E. Haff, cartographer of the official map of Oz (available from the International Wizard of Oz Club, which publishes the *Baum Bugle*, P.O. Box 26249, San Francisco, CA 94126) places the island in the Nonestic Ocean, due east of Oz. The four smaller islands of the story form a chain extending to the northeast. The isle of Phreex is mentioned on page 20 of *Rinkitink in Oz*, and in the same book page 294 also refers to the Mifkets, whose island is the second to be visited by John Dough. (The Mifkets should not be confused with the Mifkits, in John R. Neill's *Scalawagons of Oz*. Mifkits can remove their heads and hurl them at enemies. They resemble closely Baum's Scoodlers, in *The Road to Oz*.) The King of the Beavers and his subjects, who live under the waterfall on Mifket Island, reappear in Jack Snow's *The Shaggy Man of Oz*.

The ersatz General of Phreex, whose entire body has been replaced by artificial parts, raises the same perplexing metaphysical questions about personal identity as does the Tin Woodman. Ali Dubh's Elixir of Life is similar to the Powder of Life that vivifies Jack Pumpkinhead and the Saw-Horse in *The Marvelous Land of Oz*. The Beaver King's magic box anticipates Ozma's Magic Picture, which in turn foreshadows the television screen. The electrically operated ornithopter—it works so well that its inventor, Imar, is in disgrace among his fellow cranks—recalls the flying Gump of the second Oz book. As MacFall points out in his biography of Baum, the Wright brothers' flight had taken place in 1903, only a year before the second Oz book, and Baum had been quick to introduce flying ships into his stories. Two of Baum's pseudonymous books for teenage girls, *The Flying Girl* and *The Flying Girl and Her Chum*, were about a girl aviator.

The humor in *John Dough* ranges from low-level wordplay ("I'm sure I couldn't agree with anyone who ate me" John declares; I counted more than fifteen puns on words relating to food alone) to occasional remarks of existential import. John cannot recall when he was not alive. He informs a lady who thinks he ought *not* to be alive that he cannot help it. And on another occasion he observes that "it is better to be wrong than to be nothing." ("Better bread than dead," Hearn has observed.)

In 1910 the Selig Polyscope Company, Chicago, released a one-reel film of *John Dough*, starring Joseph Schrode as the gingerbread man and a girl named Grace Elder as Chick. The Selig studios had made most of the film a few years earlier for Baum's ill-fated series of "radio plays." These plays, each based on one of Baum's fantasies, had nothing to do with "radio" as the term is understood today. They were a curious mix of live actors, silent film clips (made by Selig and hand-tinted in France), colored stereopticon slides, live orchestral music, and commentary by Baum himself. Baum stood on the side of the screen with a pointer and at times moved into the film by walking off the stage and onto the screen. In 1908 Baum and one of his sons, who served as projectionist, toured fourteen cities with this remarkable show, starting in Grand Rapids and ending in New York.

The project was a financial bust. Selig obtained rights to the films, and by patching them together and adding more footage, produced four one-reelers, one of which was *John Dough*. The film seems not to have survived. In its radio play form it followed the book closely (omitting the eating of part of John by the Princess, and the visit to Pirate Island), but the movie version had a much different ending.[5] According to Richard A. Mills (see his article on the radio plays in the *Baum Bugle*, Christmas 1970), Chick somehow manages to meet with Ozma who prophecies:

> The throne of Lo-Hi shall vacant be
> Until the coming by air or sea
> Of an overbaked man and Cherub wee.

John Dough then arrives in Oz to fulfill the prophecy and be crowned king of Lo-Hi. One of the highlights of the original film (presumable also of the 1910 version) was the Fourth of July fireworks scene in which John is carried into the clouds by a giant skyrocket. In an article on the radio plays which Baum wrote for the *New York Herald* (September 26, 1909), he explains how stop-action photography was used to substitute a dummy for the live actor, just before takeoff, and to replace the dummy with the actor after the dummy falls to the ground.

It seems to me that *John Dough*, even today, could be the basis for a delightful stage or motion picture musical. Little is dated about its characters, plot, or humor. Is that smiling, long-haired Cherub, on the side of the highway with upraised thumb, a boy or a girl? Think of the fun that public relations men could have with a new, unknown child star whose sexual identity is not known to the public! And what could be more appropriate now than *John Dough*'s final moral?

Hiland and Loland, where Chick, John, and the rubber bear settle at last, are two rival cultures, flourishing side by side, each firmly convinced of its own superiority, each regarding its neighbors as uncouth barbarians. The wall that separates the tall, thin Hilanders from the short, fat Lolanders is no higher than the old Great Wall of China or the new Berlin wall, or a hundred other "walls" that these material structures symbolize. Baum's vigorous plea for tolerance and understanding of alien ways is one that he would stress again in *Sky Island*, where he describes the equally irrational rivalry between the Pinks and the Blues. Need anyone be reminded that it is a moral on which the world's fate may depend?

NOTES

1. Baum's letter is quoted by David Greene in the *Baum Bugle*, Autumn 1971, page 15.

2. There were four issues of the first edition. The first has the misprint "cage" (for "cave") on page 275, line 10. The second state corrects the error. The third (and all later printings) has no back cover picture, and "Co." is omitted after the publisher's imprint on the spine. This imprint is reset in capitals on the spine of the fourth issue.

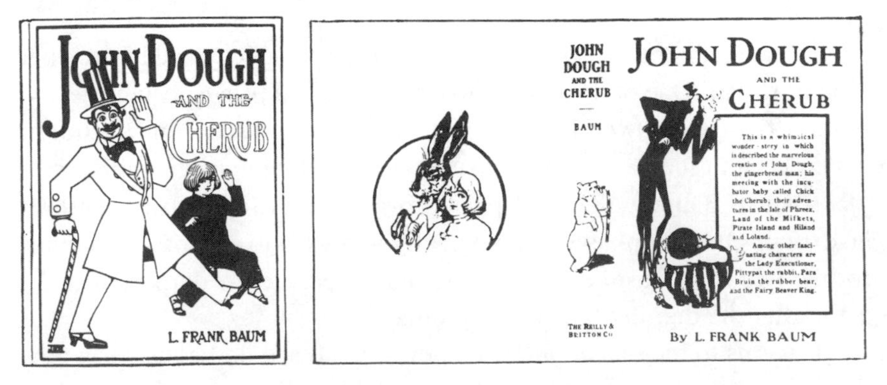

Figure 26: Front cover and dust jacket of the first issue of *John Dough*.

The second edition, published about 1920 by Reilly and Lee, eliminated color from the head pieces and most of the full-page illustrations, retaining color on twelve full-page plates. The same company's next and last edition, *circa* 1930, drops all color plates except the frontispiece. A paperback edition, newly illustrated by a young Chinese artist, Lau Shiu Fan, was published in English in Hong Kong, by Opium Books, in 1966. For additional bibliographic details see Dick Martin's report on *John Dough* in the *Baum Bugle*, Spring 1969.

A small book called *The Gingerbread Man* was published by Reilly and Britton in 1917 as one of Baum's six "Snuggle Tales" books. It reprints the first four chapters of *John Dough*, and adds a new chapter, "Safe at Last," that tells of John's arrival on the Isle of Phreex.

3. Perhaps a reader in Chicago can run down the contest results. According to the contest blank, the judges were Baum, the Chicago author Henry M. Hyde, and Wilbur D. Nesbit of the *Chicago Evening Post*. Was the book serialized in the *Post*? If so, the *Post* probably announced the outcome of the contest. The contest blank gave December 31, 1906, as the closing date, and said that prize winners would be announced about January 15, 1907.

4. Michael Hearn has noticed that John Dolittle, MD, is John Do with "little" tacked on, and that both Dr. Dolittle and John Dough were capable of speaking all bird and animal languages.

5. The only known source for the film's plot is a garbled synopsis that appeared in *Moving Picture World* sometime in December 1910.

28.

THE MAGICAL MONARCH OF MO

This first appeared as an introduction to the Dover paperback edition (1968) of Baum's fantasy about a land not far from Oz.

Russell MacFall, in his biography of Lyman Frank Baum, *To Please a Child* (Reilly & Lee, 1961), speaks of the year 1900 as Baum's *annus mirabilis*. Baum was then living in Chicago and struggling to support a wife and four sons by editing a magazine for window trimmers. His first book of stories, *Mother Goose in Prose*, had appeared three years earlier in a handsome edition published in Chicago by Way and Williams and illustrated by Maxfield Parrish. The book was not a financial success but neither was it a failure, and Baum tried again in 1899 with *Father Goose: His Book*, a collection of nonsense verse. It was published by the small Chicago firm of George M. Hill, with illustrations by the then well-known Chicago newspaper cartoonist William Wallace Denslow. For reasons that today are hard to understand, *Father Goose* was a honking success.

In April 1900, Baum wrote to his younger brother Henry Clay Baum (a throat specialist in Syracuse, New York, and the person to whom *The Magical Monarch of Mo* is dedicated):

> The money has been a pleasure to me and my work is now sought by publishers who once scorned my contributions. Harper Brothers sent a man here last week to try to make a contract for a book next year.

> Scribner's wrote offering a cash advance for a manuscript. Appletons, Lothrops and the Century have asked for a book—no matter what it is. This makes me proud, especially as my work in *Father Goose* was not good work, and I know I can do better. But I shall make no contracts with anyone till next January. If my books succeed this year I can dictate terms and choose my publishers. If they fall down I shall try to discover the fault and to turn out some better work.

The books of "this year" to which Baum referred were five children's books for which he was then already under contract. Some of the verses of *Father Goose* had been set to music, and *The Songs of Father Goose* was to be published by George Hill. The same house was also issuing two picture books of inconsequential verse by Baum, *The Army Alphabet* and *The Navy Alphabet*, and a full-length novel, illustrated by Denslow, called *The Wonderful Wizard of Oz.*

The fifth book was a collection of fourteen funny tales that Baum had written four years earlier. He had originally called it *The King of Phunnyland*, and Way and Williams had planned to issue it in 1898 as a successor to *Mother Goose in Prose*. But Way and Williams had gone out of business and the manuscript was much in need of a publisher. When the 1899 sales of *Father Goose* prompted Robert Howard Russell, a New York City publisher, to request a book from Baum, Baum produced this manuscript. Russell was delighted. William Francis Ver Beck (1858–1938), an Ohio artist who specialized in comic animal drawings (he had written and illustrated a number of children's books and later would write and illustrate many more), was asked to do the pictures for *Phunnyland*. It was published in October 1900, a month after *The Wizard* had gone on sale, under the title *A New Wonderland.*

In a first edition copy owned by Mrs. Robert S. Baum, the wife of one of Baum's sons, Baum wrote:

> This book was received by me on October 8th, 1900, although the date of publication was announced for September 20th. Mr. Russell tells me he has printed ten thousand as a first edition. I like the illustrations more than those in any of my other books up to the present time, and consider the book as a whole very pretty. It was the first children's book I ever wrote—written in 1896—but it was not offered for publication until after *Mother Goose in Prose* appeared. Then Way & Williams accepted

> it; but failed (1898) the year it was to be published, and the MS was transferred to H. S. Stone & Co., who agreed to bring it out the fall of 1898, but let it drag until too late to secure a proper illustrator. I then took the MS away from them and in the summer of '99 sent it to Russell. He accepted it too late for publication that year, so it as not issued until now. . . . I intended this book to be dedicated to my four boys, but by error the dedication was omitted. The boys heard most of these stories told before I wrote them down.

Six months earlier, in his letter to his brother, Baum had wondered which book would sell best, *The Wizard* or *A New Wonderland.* "The queer, unreliable public has not yet spoken," was the way he put it. Before the end of the year the public had spoken in a thunderous fiat. *The Wizard* was the first of fourteen Oz books destined to propel Baum into the ranks of the world's greatest writers for children.

It is easy to understand why *The Wizard* was the more successful of Baum's two 1900 fantasies. It was a single, sustained story, not a collection of short ones. Oz was a more fascinating, more imaginatively designed fairyland. Above all, it had an attractive heroine from Kansas with whom American children could identify. Eclipsed by *The Wizard*'s instant popularity, *A New Wonderland* was allowed to go out of print, but in 1903 Bobbs-Merrill reissued it under a new title, *The Surprising Adventures of the Magical Monarch of Mo and His People*—a title shortened on the cover to *The Magical Monarch of Mo.* The new title was clearly an attempt to copy the alliteration of *The Wonderful Wizard* by turning the Ws upside down, and to introduce a new and hopefully catchy two-letter word like "Oz." Baum had originally called his imaginary country the Beautiful Valley of Phunnyland. For the Bobbs-Merrill edition he rewrote the first chapter and changed the valley's name throughout the book to Mo.

It is the Bobbs-Merrill text that is used here, but the pictures have been reproduced primarily from *A New Wonderland*, which contained four color plates, counting the title page, that were not in the 1903 edition. Each edition contained text illustrations not in the other. The new text pictures in the 1903 edition have been retained in the Dover edition so that all of Ver Beck's Mo illustrations would be brought together for the first time.

Figure 27: "No wonder the Duchess Bredenbutta stared in surprise."

There are many ways in which Mo resembles Oz. It is a land of enchantment where almost anything marvelous can happen. Its human inhabitants do not grow old or die, although wild beasts can be killed. Mo animals, like those in Oz, can talk. It is a land without money or poverty. All sorts of useful things, such as neckties, swords, bicycles, and finger rings, grow on trees, as they do in certain parts of Oz as well as in Ev, a land to the west of Oz. Mo differs from Oz in two main respects: its sun never sets and it abounds in geological features made up of things children like to eat or drink—jelly mud, gumdrop pebbles, custard ponds, and so on. Baum speaks in his first chapter of "two rivers" in Mo, but actually there are four: Milk River, the River of Needles, Rootbeer River, and a river of pure maple syrup. Mo rain is lemonade. Mo snow is hot buttered popcorn.

According to the International Wizard of Oz Club's official map (prepared by cartographer James E. Haff and drawn by artist Dick Martin), Mo is a small country situated just across the Deadly Desert opposite the southeast corner of Oz where the Munchkin and Quadling regions come together. Students of Oz will remember that Mo is mentioned in two Oz books. The Wise Donkey, in *The Patchwork Girl of Oz* (page 93), reveals that he is not a native Munchkin. Indeed, he is none other than the Wise Donkey of Mo who, in this book's Fourth Surprise, digests all the books in the school library. He had been visiting Oz, he explains in *Patchwork*, when Glinda (at the close of *The Emerald City of Oz*) cast a spell over Oz that cut it off forever from the rest of the world, making it impossible for him to return to Mo.

In *The Scarecrow of Oz* (chapters 6, 7, and 8), Trot, Cap'n Bill, and the Ork stop off in Mo on their way to Jinxland. On top of a Mo mountain they visit the Bumpy Man and find Button-Bright buried under a popcorn snowdrift. From this episode we learn some unusual new facts about Mo. For example, no weather vanes are needed there. One has only to sniff. Winds from the south have a violet odor, the north breeze smells like wild roses, the east breeze like lilies-of-the-valley, and the west wind like lilac blossoms.

Oz students will also find in the Mo stories a number of places where characters and episodes call to mind similar characters and episodes in Baum's Oz books. The Cast-iron Man is a robot who precedes Tik-Tok. The Yellow Hen reappears in *Ozma of Oz* as Billina. The scene in the Cave of Daggers resembles a cave scene near the close of *Rinkitink in Oz*. And there are plot gimmicks that turn up in the fantasies of other writers. One thinks of Mary Poppins taking the children to meet Mr. Turvey (chapter 4 of *Mary Poppins Comes Back*) in a town where everything turns topsy-turvy; and a story by Lord Dunsany (one of his Joseph Jorkens travel tales) in which jungle monkeys trap an explorer in a cage.

There are several respects in which *The Magical Monarch of Mo* differs in emphasis from Baum's Oz books and his other fantasies. First of all, it is richer in humor of the Carrollian variety—humor that exploits outrageous logical impossibilities. The King escapes from a hole in the ground by pushing it over until it is upside down, thus bringing him to the surface. A tiny elephant vanishes by jumping down its own throat. The formidable Cast-iron Man is tickled by a feather and wounded by a pin. Inhabitants of Turvyland speak when they are silent, are silent when they speak, and (like the Red Queen behind Carroll's Looking-Glass) stand still when they are running. An apple on a high branch is inaccessible because the tree's trunk had been sawed off from the branch and the tree chopped up for kindling wood. A Wizard has morning office hours from ten forty-five to a quarter to eleven. Perhaps it was this Carrollian nonsense that led Baum, in his first book title, to speak of Mo as a "new" wonderland.

Mo also is richer in puns than most (not all) of Baum's other books. The King, wearing his wooden head, is hardheaded. A dog is something of a wag. Sore throats are "cured" by hanging them out to dry in the sun. Names of characters are often puns: the King's royal chamberlain Nuphsed (enough said), Duchess Bredenbutta (bread and butter), the giant Hartilaf (hearty laugh), and others. Baum, like Carroll, always enjoyed this sort of wordplay but there is a bit more of it in *Mo* than usual.

Finally, *Mo* is richer in the kind of sadistic absurdities that later became so characteristic of the animated cartoons of Walt Disney and his competitors. I refer to such things as severed heads, a bitten-off toe, scratched-out eyes, a dog flattened to a pancake, a Prince who goes through a giant's clothes wringer and is restored by being pumped up through a hole in his head, and a Purple Dragon destroyed by stretching him so thin that he can be cut up into fiddle strings with excellent tones. It is all done in such an offhand, preposterous way that it is hard to imagine how children could be disturbed by it, but evidently Baum himself later decided that this kind of comedy is best soft-pedaled. There is much less of it in most of his other books.

At the beginning of his charming introduction Baum writes that he is not ashamed to admit that he is still a child at heart. In the letter to his brother quoted earlier, he puts it this way: "The boys are growing wonderfully and I sometimes think I must be a kid no longer, when I behold the stalwarts around me and hear them call me 'dad.' There's a mistake somewhere, for I have failed to grow up—and we're just five boys together. . . ."

Was there ever a popular writer of fiction for children who did not, in certain respects, forget to grow up? It was surely one of Baum's great trade secrets. Having had the privilege of meeting two of his sons, I can testify that no sons ever loved their father more. And I suspect that, although a dwindling number of librarians still cannot quite bring themselves to admit it, few American children have ever loved a writer more.

29.

AMERICAN FAIRY TALES

This was my introduction to Baum's book of short unrelated fantasies when Dover reprinted it as a paperback in 1978.

The title of this book is something of a misnomer. Not all its stories occur in America, and not all are about fairies. Most of them, however, have native settings, and all are "fairy tales" in the broader sense of being fantasies. In a preface to a later edition, Lyman Frank Baum maintained that never before had a book contained "fairy tales" that take place, for the most part, in the United States. I do not know if this is true or not. But if there were earlier books of this sort for children, they were by now-forgotten writers.

In 1901 when *American Fairy Tales* first went on sale, Baum had a well-established reputation. Two of his books, *Father Goose: His Book* (1899) and *The Wonderful Wizard of Oz* (1900) had been excellent sellers. Both were preceded by a less successful collection of short fantasies, *Mother Goose in Prose* (1897), and in 1900 another book of short stories, *A New Wonderland*, had been published. (The second book was later retitled *The Magical Monarch of Mo*). It is true that Dorothy and Toto came from Kansas, and the Wizard from Omaha; but their adventures take place in Oz, all the characters of *The Magical Monarch* live in Mo, and most of the settings of the Mother Goose tales are in England, as they should be. None of the stories in this last collection is clearly identified as taking place in this country, although the final story, about a little farm girl named Dorothy, is surely American because it speaks of Santa Claus instead of Father Christmas.

The twelve stories of *American Fairy Tales* were first syndicated weekly in at least five newspapers during the first half of 1901. Four of these papers are named by the newsboy on page 173 when he calls out: "Chronicle, 'Quirer, R'public 'n' 'Spatch! Wot'll ye 'ave?" (the Chicago *Chronicle*, Cincinnati *Enquirer*, St. Louis *Republic*, and Pittsburgh *Dispatch*). The stories also ran in the Boston *Post*. The book's illustrations by Ike Morgan, Harry Kennedy, N. P. Hall, and Ralph Fletcher Seymour (who did the cover, title page, and border pictures) were the basis of most of the drawings that accompanied the stories in the *Enquirer*, *Dispatch*, and *Post*. Pictures in the *Chronicle* and the *Republic* by staff artists were, I am told, quite different.

George M. Hill, who handled newspaper syndication rights and printed the book's first edition in 1901, was the small Chicago house that had earlier published *Father Goose* and *Wizard*. In the same year that *American Fairy Tales* appeared, Hill also issued Baum's second full-length fantasy, *Dot and Tot of Merryland*. It would be several years before Baum's readers, and the great success of the 1902 stage version of *Wizard*, persuaded him to return to Oz.

A completely new edition of *American Fairy Tales* was published by Bobbs-Merrill in 1908. All the illustrations from the Hill edition were replaced with sixteen two-color plates by George Kerr, a cartoonist for the New York *Herald*. The twelve stories were rearranged—for the worse, because the weakest story, "The King of the Polar Bears," was shifted to the end, thus destroying the lovely ending of the first edition. All the stories were revised in numerous minor ways, and three entirely new tales were added. Baum also added the following "Author's Note":

> This is the first time, I believe, that a book has been printed containing Fairy Tales that relate mainly to American boys and girls and their adventures with real fairies in the United States and other American countries.
>
> If fairies exist at all—and no one has yet been able to prove that they do *not* exist—then there is no good reason why they should not inhabit our favored land as well as the forest glades and flowery dales of the older world across the water. For fairies are not peculiar to any one locality, and every race has its own fairy legends.

Figure 28: The frontispiece illustration by George Kerr from the second (1908) edition of *American Fairy Tales*.

Yet, we must consider that the beautiful and well-known tales of Andersen and the Brothers Grimm, as well as those of Hauff, Perault [*sic*], Caballero and Andrew Lang, date many long years ago, and such histories would never do for American Fairy Tales, because our country has no great age to boast of. So I am obliged to offer our wide-awake youngsters modern tales about modern fairies, and while my humble efforts must not be compared with the classic stories of my masters, they at least bear the stamp of our own times and depict the progressive fairies of to-day.

My friends, the children, will find these stories quite as astonishing as if they had been written hundreds of years ago, for ours is the age of astonishing things. They are not too serious in purpose, but aim to amuse and entertain, yet I trust the more thoughtful of my readers will find a wholesome lesson hidden beneath each extravagant notion and humorous incident.

L. Frank Baum
Macatawa, 1908

Sometime around 1924 Bobbs-Merrill issued a third edition with a new pictorial label on the cover and Kerr's color plates reduced to eight. It is not known how often this edition was reprinted, but it was in print until 1942. Four stories from the 1901 edition are in *The Purple Dragon and Other Fantasies*, a collection of Baum short stories edited and annotated by David L. Greene (Fictioneer Books, Ltd., 1976).

Children seemed not to care much for *American Fairy Tales*, though it is hard to tell because only grown-ups buy children's books and write reviews. Is fantasy at its best only when it takes place in a far-off region or long ago? J. M. Barrie wrote a dull book about Peter Pan in Kensington Gardens; it was only when Peter took Wendy to Neverland that Peter became famous. Of course there are exceptions (Mary Poppins!). Classics such as *The Wind in the Willows* and *Charlotte's Web* are not exceptions, because talking animals alone do not a fantasy make—they are merely devices for disguising real people.

"*American Fairy Tales*, I am sorry to tell you," wrote James Thurber in a short essay on Baum, "are not good fairy tales. The scene of the first one is the attic of a house 'on Prairie Avenue, in Chicago.' It never leaves there for any wondrous, far-away realm." Russel B. Nye, in an appreciation of Baum, elaborates the same point:

> The *American Fairy Tales* were good stories, far better than most run-of-the-mill "educational" tales for children, but in the majority of them Baum failed to observe the first rule of the wonder-tale—that it must

> create a never-never land in which all laws of probability may be credibly contravened or suspended. When in the first story the little girl (Dorothy by another name) replies to a puzzled, lost genie, "You are on Prairie Avenue in Chicago," the heart goes out of the story. It is only in Quok, or in Baum's zany version of the African Congo, or among the Ryls, that the book captures the fine free spirit of Oz. The child could see Chicago (or a city much like it) with his eyes; Oz he could see much more distinctly and believably with his imagination. Baum nevertheless clung for a few years to the belief that he could make the United States an authentic fairyland. "There's lots of magic in all nature," he remarked in *Tik-Tok of Oz*, "and you may see it as well in the United States, where you and I once lived, as you can here." But children could not. They saw magic only in Oz, which never was nor could be Chicago or Omaha or California or Kansas.

To this criticism we may add that Baum injected into the stories more than his usual quota of adult farce—witticisms that few young readers could appreciate. The Women's Anti-Gambling League sponsors a weekly card party. An Italian bandit, tightly packed for a long time in a chest with two of his comrades, refers to one of them as his "nearest friend for years." When one of the bandits tells Martha that his business is to rob, the girl informs him that jobs with the gas company are hard to get, but he might consider becoming a politician. A wealthy crone, who is the highest bidder for the right to marry the ten-year-old King of Quok, pays $3,900,624.16 "in cash and on the spot, which proves this is a fairy story." As Michael Hearn points out, only adult readers would perceive this story as a satire on a common practice of the time, the marriage of wealthy American women to younger and destitute Europeans with royal titles.

In spite of Baum's cynical attitude (his irrelevant "morals" at the end of many of the stores are obviously tongue-in-cheek), most of the stories are excellent yarns, and alert children can appreciate much of the humor. Particularly amusing are the dialog about the pie that got packed into the chest with the bandits; the senator's compulsive ballet dance after he has just declared it to be "a most impressive and important occasion"; the merry laugh ("Guk-uk-uk-uk!") of Gouie, the hippopotamus, and the surprise ending of the story about him; and the dress that is reduced from $20 to the great bargain of "only $19.98." Judging by its many reprints, "The

Magic Bon Bons" was the book's most popular story. It was even made into a movie by Baum's Oz Film Company.

Science fiction buffs will find "The Capture of Father Time" of special interest. There have been endless stories about time speeding up or slowing down—H. G. Wells's "The New Accelerator," for example—but stories about the world's gyrations coming to a total halt (except for the observer) are not common. Baum handles the theme skillfully. "The Wonderful Pump" is a well-constructed imitation of a traditional fairy tale. And who can read "The Enchanted Types," a biting satire on female fashions—especially the outlandish turn-of-the-century fad of wearing stuffed real birds on hats—without instantly becoming a disciple of Thorstein Veblen?

Baum enthusiasts will find many parallels with ideas in other Baum books. The notion of a glass animal was used again in *The Patchwork Girl of Oz* when Dr. Pipt brings to life a transparent glass cat named Bungle. The butterfly, who has such a tender heart but no soul or conscience, recalls the Tin Woodman of *Wizard* taking great pains to be kind because he lacks a heart.

Knooks and Ryls inhabit the Forest of Burzee in *The Life and Adventures of Santa Claus* (now also back in print as a Dover edition), and along with Santa they attend Ozma's birthday party at the close of *The Road to Oz*. A Ryl is prominent in three other Baum stories. One is "The Ryl," a tale added to the 1908 edition of *American Fairy Tales*. Another, "The Runaway Shadows," was a newspaper story intended to be part of the 1901 serialization, although it did not appear until much later. It was reprinted for the first time in the *Baum Bugle* for April 1962. Tanko-Mankie, the Yellow Ryl in "The Dummy That Lived," reappears in "The Yellow Ryl," a story apparently first published in two parts in the August and September 1925 issues of a children's magazine called *A Child's Garden*. The story was perhaps intended for the 1908 edition of *American Fairy Tales*. The *Baum Bugle* reprinted it in 1964 in its Spring and Autumn numbers.

Although Baum's style is sometimes pedestrian, his narratives move along briskly, seldom flagging in interest and Ozzy invention. There are spots where his sentences have a spare, concise beauty. "He tossed and tumbled around upon his hard bed until the moonlight came in at the window and lay like a great white sheet upon the bare floor." The book's final sentence startles with the abruptness, brevity, and elemental poetry of a *haiku*.

> The butterfly flew away to a brook
> and washed from its feet all traces
> of the magic compound. When night
> came it slept in a rose bush.

When Jim releases Father Time, "with a rustle and rumble and roar of activity the world came to life again and jogged along as it always had before." This is Baum writing with his usual carelessness, but creating vivid images that jog his tale along for the delight of all of us, old and young, who have fallen under the Royal Historian's magic spell.

30.

MOTHER GOOSE IN PROSE

This review first appeared in the *Baum Bugle*, Winter 1997.

In 1991, Bette Goldstone, an education professor at Beaver College, Glenside, Pennsylvania, completed a survey of 150 preschool children in suburban Philadelphia to determine how much they knew about Mother Goose rhymes. More than a third had never heard of "Jack Be Nimble," "Hey Diddle Diddle," or "Little Miss Muffet," nor could they say what happened to Jack and Jill. "We're not losing the fairy tales," Goldstone said in an Associated Press interview that ran in newspapers for May 2, 1991. "They've all been converted to cartoons and motion pictures. But we are losing Mother Goose."

This decline of interest in the old nursery rhymes may be one reason why L. Frank Baum's *Mother Goose in Prose* has been so completely overlooked by today's critics of juvenile literature. At the time Baum wrote his book, almost every child in America could recite a dozen of the classic rhymes. When I was a child in the 1920s, my mother framed half a dozen reproductions of Mother Goose paintings by Jessie Willcox Smith and hung them in a row on the wall of the nursery where my siblings and I played. I doubt if any of my children or grandchildren know a single Mother Goose rhyme from beginning to end.

Yet somehow the old rhymes live on. In *Book World* (October 31, 1965), Maurice Sendak reviewed eighteen books about Mother Goose that had recently been published. Of course, only adults buy books, and I suspect that few children responded favorably to any of the reviewed volumes. Sendak was particularly harsh on the art in Brian Wildsmith's highly touted Mother Goose. He found the illustrations "noisy, posturing" and "cinematic *Mother Goose* gone Wildsmith." The verses were scrunched down at the bottom of each page, Sendak said, implying "only too clearly their unimportance."

Baum was not the first to weave fictional events around familiar nursery rhymes. Lewis Carroll, in his two *Alice* books, provided contexts for the Jack of Hearts who stole the tarts, for the battle of the Tweedle brothers, for the fight between the Lion and the Unicorn, and for Humpty Dumpty's immortal fall. Whether this suggested to Baum that he do the same with twenty-two nursery rhymes, I do not know. We do know that Baum knew the *Alice* books well.

In 1896, Baum was living in Chicago, barely earning a living for his family by working as a traveling salesman for a china and glassware company. He had earlier published only one book, *The Book of the Hamburgs*, on how to raise and care for this breed of chickens. But he had started selling adult short stories to different magazines, and was beginning to believe that he could become a profitable freelance writer. He worked hard on two books of fantasy for children. One would eventually be titled *The Magical Monarch of Mo*. The other was a book he first called *Tales From Mother Goose*.

Although *Mother Goose in Prose* was Baum's first published book for children, it was not the first such book that he wrote. We know from an undated interview with Baum (circa 1913), recently discovered by Michael Patrick Hearn, that the Mo stories were written first. Baum originally called the magic region Phunniland, and when the book was published in 1900 it had the title *A New Wonderland*. Hearn has also reminded me that in Baum's own copy of *A New Wonderland*, he noted, among other things: "Written in 1896. . . . It was the first children's book I ever wrote, but was not offered for publication until after *Mother Goose in Prose* appeared."

Mother Goose in Prose was published by the Chicago firm of Way and Williams in 1897. It was the first book illustrated by Maxfield Parrish although, as Hearn informs me, Parrish had previously done the cover for the Way and Williams edition of Opie Read's *Bolanyo*. Parrish was twenty-seven in 1896, and destined to become one of our nation's most admired painters and graphic artists.

According to Hearn, the Way and Williams edition of *Mother Goose in Prose* did not sell well, perhaps because of its relatively high price, for the time, of $2.00. The publisher went bankrupt in 1898. Herbert S. Stone and Company, in Chicago, took over the stock, Hearn tells me, but they did not reprint the book. Rights reverted to Baum, and he sold the book to George M. Hill, another Chicago house, for the smaller, less expensive second edition.

Critics have faulted the stories in *Mother Goose in Prose* for being dull and sentimental. This may be true of some of them, but for the most part I find them well written, delightful, and rich in humor. I doubt if many writers could have conceived such ingenious explanations of how two dozen blackbirds got baked in a pie, how a cow managed to jump over the moon, how a bramble bush could scratch out a man's eyes and another bush scratch them back again, or how an old woman's house came to resemble a shoe.

Who but Baum would have thought to name three mice Hickory, Dickory, and Dock? As for the sentimentality, who said that children dislike sentiment in stories? I confess unashamedly that when as a child I first read "The Jolly Miller," with its eternal theme that everybody loves somebody sometimes, it brought tears to my eyes.

There are many foreshadowings of Oz in Baum's Mother Goose tales. The very first story begins with a boy named Gilligren, an orphan like Dorothy, walking a dusty road on his way to see, not a wizard, but a king. Baum well knew, as so many writers of juvenile fiction do not, that the surest way to grab a child's interest is to start a story with a boy or girl heading down an unfamiliar road toward unpredictable adventures. And does not the boy's name suggest the purple Gillikin region of Oz, so called (as I have conjectured before) after the name of the purple gilly flower that flourished around Baum's childhood home?

Although not living in Oz, many of Baum's Mother Goose animals can talk, starting with sheep in the fourth tale and ending with Bun Rabbit in the last. Old King Cole, fat, jolly, and riding a donkey, surely prefigures fat and jolly King Rinkitink on his faithful goat. The Man in the Moon is sent home, like the Wizard, in a balloon. The Wise Men of Gotham are fully aware, as was the Wizard in the first Oz book, that they are "arrant humbugs."

The method of choosing a king at random, in King Cole's story, is repeated in *Queen Zixi of Ix*, and also, let me add, in G. K. Chesterton's fantasy *The Napoleon of Notting Hill.* The contrariness of laws on the moon, where hot is cold and cold is hot, suggests Mo's Turvyland, where everything is upside down, and the boy in *The Enchanted Island of Yew* who enjoys pain. The magic collar that protects Bun Rabbit prefigures the magic pearl that protects Inga in *Rinkitink in Oz*. Like Trot and Button-Bright, Miss Muffet does not miss her parents when she is off on an adventure.

England, with its colorful heritage of castles and royalty that enter into so many old nursery rhymes, naturally is reflected in Baum's stories, as well as in so many of his full-length fantasy novels. All the towns and regions mentioned in the tales are British, including Cumberland, a county near Scotland in northern England. The one exception is Whatland in the King Cole story. Does it belong on the map of regions surrounding Oz?

As all Baum buffs know, the most important anticipation of Oz occurs in the final tale where we meet a farm girl named Dorothy who has the ability to talk to animals. When Baum later included this story in his *Juvenile Speaker* and in the *Snuggle Tales* series, he changed her name to Doris to avoid confusing her with Dorothy Gale of Kansas.

Unlike so many European fairy tales, almost all of Baum's Mother Goose stories have happy endings. One exception is the tragic account of the fates of Humpty Dumpty and his brown egg girlfriend Coutchie-Coulou. It is certainly a strange name for an egg. In Baum's day, carnivals and circuses featured a dance called the "hootchie cootchie" in which young women wobbled their hips like an egg when you spin it on end. And was "Coulou" intended to echo the exclamation "callooh!" in Lewis Carroll's "Jabberwocky," a nonsense poem explicated by Humpty from his perch on the wall before he fell?

Michael Patrick Hearn reminds me that parents like to say "cootchie-coo" when they tickle babies, but whether this practice came before or after the dance I don't know. In *Father Goose, His Book* (1899), illustrated by William Wallace Denslow, Baum introduces Coutchie again, this time as a brown girl from India in his nursery rhyme that goes:

Cootchie Cooloo
    Was a Girl of Hindoo

Who was rather too
large for her size;
Her teeth were quite white
And her nose was all right,
But she had a bad squint
to her eyes.

The other exception to a happy ending is the drowning of the three Wise Men of Gotham. Incidentally, in the first edition of this book, the illustration has the Wise Men's tiny boat on a slant to portray the perils of high seas. Baum must have thought his child readers would not identify with eggs, and so not mind the horrible deaths of Humpty and his friend. In this he was surely mistaken. Even the Wise Men are harmless, likable frauds who seem not to deserve being lost at sea.

More than many juvenile writers, Baum realized that children are bored by long passages about the wonders of nature.* However, when he devoted a sentence or two to such things, his prose has a simple, elemental, poetic charm. "The world was before him, and the bright sunshine glorified the dusty road and lightened the tips of the dark green hedges that bordered the path" (page 21), "listening to the singing of the birds, and the gentle tinkling of the bells upon the whethers, and the far-away songs of the reapers that the breeze brought to his ears" (page 38). "One spring, just as the grasses began to grow green upon the cliff and the trees were dressing their stiff, barren branches in robes of delicate foliage" (page 75). "The little brook that wandered through the meadows, flowing over the pebbles with a soft, gurgling sound that was very nearly as sweet as music" (page 85).

Subtle touches are easily missed on a first reading, such as the wealthy mother in "Little Miss Muffett" who "happened to be home for an hour" when she suddenly remembered she had a daughter in the house. And there is the sly doctor, in the same tale, who, when he discovers his early diagnosis was correct and recalls how rich the family is, jacks up his fee.

---

*Ruth Berman called my attention to Baum's comment on the "beautiful descriptive passages" in Hans Christian Andersen's fairy tales. Writing on "Modern Fairy Tales," in *Advance*, August 19, 1909 (reprinted in *The Wizard of Oz*, Schocken, 1983, edited by Michael Patrick Hearn), Baum had this to say: "As children you skipped those passages—I can guess that, because as a child I skipped them myself. . . ."

Baum's droll sense of humor is everywhere, especially in the reactions of slow-witted people to the Wondrous Wise Man's preposterous riddles. Carrollian logical nonsense merges most clearly in the last tale when Bun Rabbit, seeing how lifelike is the rabbit made by Santa, can't decide whether he is the real rabbit or the fabrication.

Most of the Mother Goose stories are not fantasies; talking animals are a device, not full-blown fantasy. The first real fantasy is the tale about the Man in the Moon. In his second sentence, Baum admits that although his story is amusing, "there is not a word of truth in it." There also is not a word of truth about the moon in a passage in "The Cat and the Fiddle." Baum tells us that "the sun was sinking behind the edge of the forest and the new moon rising in the east." This mistake is often made by writers unfamiliar with astronomy. When the moon rises at the same time the sun sets, it is always a full moon, never a crescent.

Baum was a mediocre poet, but he loved to write doggerel. There is lots of it in this book, starting with the verses about Mistress Mary. One tale, "Pussy-Cat Mew," is entirely in verse! Tommy Tucker was surely right when, commenting on one of Baum's songs, he says: "I can't say much for the air, nor yet for the words, but it was not so bad as it might have been."

Adult economics and politics enter the book's story about the beggars who came to town. It begins as if it might be a defense of socialism, but ends with a hymn to capitalism and free enterprise that would please any of today's conservative Republicans.

I am not familiar enough with the history of nursery rhymes to evaluate the accuracy of Baum's introduction. I never heard the conjecture that Shelley might have written "Pussy-Cat Mew" or that Swift wrote "Little Bo Peep." Both Michael Patrick Hearn and I suspect that Baum intended both those conjectures to be jokes. If you wish to know more about the confusing history of Mother Goose, there are two classic works to consult: *The Oxford Dictionary of Nursery Rhymes*, edited by Iona and Peter Opie, and *The Annotated Mother Goose*, edited by William and Ceil Baring-Gould.

Every nursery rhyme used by Baum is in the two books cited above except for the one about Little Bun Rabbit. I have been unsuccessful in tracking this down. It cannot be English because it mentions Santa Claus.

It seems almost certain that Baum made it up as an excuse for a story about Santa. Later, Baum wrote a book of original nursery rhymes supposedly composed by "Father Goose," as well as an entire novel about the life of Santa Claus.

The first edition of *Mother Goose in Prose* was a strikingly beautiful gray cloth volume published in 1897 by the Chicago house of Way and Williams. Parrish's superb black-and-white illustrations were on tipped-in plates. His cover showed a boy and a girl seated facing one another. Two states are known, the second having two additional blank pages at the end. Two years later an identical book, except for a change of cover and the date of the introduction, was published in London by Buckworth and Company.

In December 1897, Way and Williams issued twenty-seven numbered copies of a deluxe portfolio of the Parrish art, each proof autographed in pencil by the artist. The set is now extremely rare. In 1900, the Pettijohn Breakfast Food Company published twelve of Baum's stories in pamphlet form. A child could obtain a pamphlet by cutting pictures of three bears from the cereal box and sending them to the company with eight cents to cover postage.

The book's second edition (1901), by Chicago's George M. Hill Company, was reduced in size and the pictures printed in sepia. Its red cloth cover shows a row of geese at the top and another row at the bottom.

Bobbs-Merrill Company of Indianapolis issued the third edition in 1905. Its first state continues the art's sepia colors. A brown cloth cover has a large goose on the right and left sides. The second state changed the cover art to a paper label reproducing Parrish's picture of the Little Man with the gun. All illustrations, including the one on the brown cloth cover, are now red on a pale yellow background. The third state continues the red color, but now the paper label, on a green cloth cover, is in full color. One assumes that Parrish did the coloring. The first edition had no captions on the art. They were added in the second and all later editions.

The fourth edition, issued in 1974 by Bounty Books, was available in a second printing in 1986, distributed by Outlet Book Company. Both these printings have the art in black-and-white. In July 1956, *Children's Digest* reprinted "The Cat and the Fiddle." (For more details about early editions, consult the *Baum Bugle*, Spring 1966 and Autumn 1981.)

When the first edition came off the press, Baum sent a copy to his sister, Mary Louise, with the following memorable inscription. It has been quoted often, but deserves quoting again:

> My Dear Mary:
>
> When I was young I longed to write a good novel that should win me fame. Now that I am getting old my first book is written to amuse children. For, aside from my evident inability to do anything "great," I have learned to regard fame as a will-o-the-wisp which, when caught, is not worth the possession, but to please a child is a sweet and lovely thing that warms one's heart and brings its own reward. I hope my book will succeed in that way—that the children will like it. You and I have inherited much the same temperament and literary taste and I know you will not despise these simple tales, but will understand me and accord me your full sympathy.
>
> Lovingly, your brother Frank.

31.

HOW THE OZ CLUB STARTED

This is scheduled to appear in the *Baum Bugle.*

Was it almost fifty years ago? I'm now ninety-three, so I must have been in my early forties when the International Wizard of Oz Club was founded in 1957. I was then living in an apartment on Charles Street in New York's Greenwich Village. My two-part essay on L. Frank Baum, titled "The Royal Historian of Oz," had appeared in the *Magazine of Fantasy and Science Fiction* (January and February 1955), then edited by Anthony Boucher, a devoted Ozian.

At that time the country's librarians, almost to a man and woman, were dismissing Baum's fantasics as not only mediocre, but unwholesome for children. Caroline Meig's 624-page *Critical History of Children's Literature* (1953) made no mention of Baum! The head of Detroit's public libraries declared that Baum's Oz books had "no value" for today's youngsters. He expressed his pride in the fact that no Oz book was allowed in any city library!

Appalled by such an opinion, Michigan University Press decided to publish a new edition of *The Wizard of Oz.* It included my biography of Baum and an essay by Russell B. Nye, chairman of Michigan State University's English Department.

At the back of that book I listed, apparently for the first time, all of Baum's many books, including those he wrote under other names. All over the land dealers in used books suddenly realized they had in stock copies of books they had no idea were written by Baum. A few years earlier, visiting a bookstore in New Jersey, I had picked up a complete set of all the books for girls that Baum had written under the pseudonym of Edith Van Dyne. They were twenty-five cents a copy. Serious collectors of Baum began to proliferate. First editions of *The Wizard*, previously selling for a few hundred dollars, started to rise in price, first to five or six hundred dollars, then eventually into the thousands.

My article on Baum quickly led to my meeting three Oz enthusiasts who lived in New York City: Jack Snow, Roland Baughman, and Justin Schiller. Jack was then working at a dreary desk in the city's Blue Cross office. Desperately in need of money, he had sold his entire Baum collection, much of it going to Baughman, who for many years had been acquiring a great Baum collection for Columbia University's library where he worked.

It was Jack who introduced me to Justin, then in his early teens. With the help of his father, an antiques dealer, he had started what would eventually become one of the nation's largest Baum collections. Justin now owns a rare books store in Kingston, New York, that specializes in early juvenile literature. His lavish catalogs are now collector items.

In 1955, to the amazement and chagrin of anti-Oz librarians, Columbia University sponsored the nation's first comprehensive exhibit of Baum's writings. I visited the show with Snow and Justin. I remember the sadness in Jack's eyes as he looked at so many of the rare first editions he had once owned. He had written two Oz books of his own. His *Who's Who in Oz* listed alphabetically all the major characters in Oz books by both Baum and Ruth Plumly Thompson, along with outlines of the plots of each book.

After Jack's death from an illness, his brother came to Manhattan to arrange for Jack's body to be taken back to his hometown for burial. There was a short obit which I had phoned to the *New York Times*. I had a hand in disposing of the few books and other Oz items that were in Jack's disheveled single-room apartment. An original watercolor of the Emerald City, by John R. Neil, was bought by Fred Meyer along with other odds and ends. On the room's wall was a huge book poster for the *Wizard*—it is reproduced on the cover of Michigan State's *The Wizard of Oz and Who He Was*. Jack's brother took the poster back with him. I don't know what happened to it. I've been told the only other known copy is in the Library of Congress.

It was Jack who told me about Fred Meyer and his great interest in all things Oz. He and Justin began corresponding, and in 1957 Justin decided it was time to start an Oz organization. He named it the International Wizard of Oz Club, and began distributing its official journal, the *Baum Bugle*. It was a mimeographed document of several pages stapled together. I wouldn't be surprised if I were the only owner of a complete run of the *Bugle* which includes all the mimeographed issues. On the masthead of the first issue you'll find me listed as "chairman of the board"!

I cannot recall who organized the club's first convention, but it was probably Fred. I missed the first gathering, but attended the second and one or two later ones. I recall being impressed by how excited the attendees were to make new friends and discover they were not alone in their love of Oz. At one convention I donated to the auction a copy of *The Woggle-Bug Book*, which had been given to me by writer and friend Everett Bleiler, who said he had picked it up somewhere for just a few dollars. I recall my amazement when someone bought it for a few hundred!

I was greatly saddened by Fred's recent death after a long illness. We had become friends, speaking often by phone. I wish I had preserved all of his Ozzy Christmas cards. He probably knew as much about Oz as anyone except Michael Hearn and Dick Martin. I first met Dick when he was in his teens and already a dedicated Ozian. I remember him telling me he had found a way of getting rid of pesky young Oz buffs who insisted on long chats with him at conventions. He would spread his hands and say "me no speak Oz."

In my introduction to Norton's handsome new edition of Hearn's *Annotated Wizard of Oz* I tell how this book came about. After the success of my *Annotated Alice*, its publisher Clarkson Potter asked if I could annotate *The Wizard*. I declined because I knew someone who could do a much better job, namely, Michael Patrick Hearn, then a student at Bard College. Clark at once visited Bard where he hired Hearn on the spot. Years later I introduced Hearn to Robert Weil, my editor at Norton who had just published a handsome updated edition of my *Annotated Alice*. He followed it with a similar updating of Hearn's *Wizard* which Crown had let go out of print. Michael has also annotated several other classics, written a biography of Denslow, and is currently working on a definitive biography of Baum.

The first biography of Baum, *To Please a Child* (1961), was written by Russell P. MacFall, then a night editor at the *Chicago Tribune*. One night, while conversing with some of his associates, MacFall said he would like to do a life of someone of importance about whom no one had yet published a book. The Judy Garland movie was then having a big success. Someone suggested that the author of the book on which the film was based might be a good candidate for a biography. No one at the table could remember the author's name!

In his life of Baum, MacFall had the assistance of Baum's oldest son, Colonel Frank Baum. The colonel visited me once when I lived in the Village. He was bitter toward his mother for not recommending him to Reilly and Lee as the person who could continue the Oz series after his father died. Colonel Baum did actually write one Oz book. A tiny volume called *The Laughing Dragon of Oz*, it sold for ten cents in the dime stores. You need only to read a few pages to realize that Maud Baum made a wise decision.

Many Oz books have been written by others. Three were by Neil, the "Royal Illustrator of Oz." For decades I resisted the impulse to write one myself because of my respect for Baum and the "cannon." I finally gave in and perpetrated a semi-satire for adults, not for children, and called it *Visitors from Oz*. The *New York Times* called it a "poor thing of a novel," but the *London Times Literary Supplement*, to my surprise, gave it a long, wonderful review. I managed to place Lewis Carroll's Wonderland in the Gillikin region of Oz, not far from a new Mount Olympus inhabited by the exiled Greek gods. I also disclosed, for the first time, that Mary Poppins lives in Oz.

The bitter controversy over Baum's reputation has happily evaporated. Anti-Baum librarians, who I once called "gray-minded" (not gray-haired!), are now in full retreat. Even British critics, thanks to the efforts of Angelica Carpenter and a few others, are discovering that there is much more to learn about Oz than you can get from the MGM musical.

PART FOUR: LEWIS CARROLL

32.

SYLVIE AND BRUNO

It was not until my late twenties that I discovered the greatness of Lewis Carroll's novels about Alice's adventures in Wonderland and behind the looking-glass, and how many interests I shared with the author. Like Carroll I am fond of recreational mathematics and logic, of puzzles, wordplay, magic, chess, and even aspects of theology.

The more I explored books about Carroll, the more I became convinced that his *Alice* books swarmed with subtleties, jokes, and literary allusions that could be grasped by Victorian British readers but would be completely missed by today's American children. In brief, neither book could be fully appreciated without the aid of many notes. I approached editors of several top publishing houses with the suggestion that they ask Bertrand Russell, known to be a great admirer of Carroll, to annotate the two *Alice* books. One editor actually wrote to Russell with the proposal, but he declined. It was Clarkson Potter, then a young editor at Crown, who suggested I do the job myself. The rest is history.

In later years I provided introductions to other books by Carroll, five of which are here reprinted. What follows is my introduction to Dover's 1988 edition of Carroll's almost forgotten fantasy *Sylvie and Bruno*.

Ever drifting down the stream—

Lingering in the golden gleam—

Life, what is it but a dream?

The lines above are the final stanza of the poem that closes the second *Alice* book, *Through the Looking-Glass*. The first letters of each line of the poem spell the name of Alice Pleasance Liddell, Lewis Carroll's most-loved child-friend. The prefatory poem of *Sylvie and Bruno* has the same meter and rhyme scheme, and the last words of its first stanza—"dream," "gleam," "stream"—are the same as the last words of the above lines except that their order is reversed.[1]

Carroll's prefatory poem to *Sylvie and Bruno* is even more remarkable as an acrostic than the earlier poem. Not only do its first letters spell Isa Bowman, perhaps Carroll's second-favorite child, but the first three letters of each stanza are "Isa," "Bow," and "Man." Isa played Alice on the stage, and later wrote a delightful book, *Lewis Carroll as I Knew Him*. When her "uncle" (as she called Carroll) sent her an inscribed copy of *Sylvie and Bruno*, Isa did not realize it was dedicated to her. "She was so long, without finding it out," Carroll wrote to a friend, "that I've had to give her a hint: and I don't *yet* know whether she has found out that it comes in *two* different ways."

Although Carroll clearly intended to link *Sylvie and Bruno* to his earlier masterwork, he also wanted it to be entirely different. The *Alice* volumes were pure fun, Carrollian nonsense without morals. Devout Anglican that he was, Carroll longed to write a more serious book in which he could combine humor with his deepest beliefs, both sacred and secular. Here is how he put it in a memorable letter:[2]

> In *Sylvie and Bruno* I took courage to introduce, what I had entirely avoided in the two *Alice* books, some reference to subjects which are after all the *only* subjects of real interest in life; subjects which are so intimately bound up with every topic of human interest, that it needs more effort to avoid them than to touch on them. And I felt that such a book was more suitable to a clerical writer than one of mere fun.
>
> I hope I have not offended many (evidently I have not offended *you*) by putting scenes of mere fun, and talk about God, into the same book.

> Only *one* of all my correspondents ever guessed there was more to come of the book. She was a child, personally unknown to me, who wrote to "Lewis Carroll" a sweet letter about the book, in which she said "I'm so glad it hasn't got a regular wind-up, as it shows there is more to come!"
>
> There is indeed "more to come." When I came to piece together the mass of accumulated material, I found it was quite double what could be put into one volume. So I divided it in the middle: and I hope to bring out Sylvie and Bruno Concluded next Christmas—if, that is, my heavenly Master gives me the time, and the strength, for the task: but I am nearly 60, and have no right to count on years to come.

The two volumes of *Sylvie and Bruno*, together running to some 850 pages, was Carroll's last major work. "Whether it is better, or worse, than the *Alice* books," he wrote in another letter, "I have no idea; but I take a far deeper interest in it, as having tried to put more real *thought* into it."

Carroll's great nonsense ballad *The Hunting of the Snark* was written in a bizarre backward way. After its last line, "For the Snark *was* a Boojum, you see," occurred to him suddenly during a walk, he composed over the next two years a ballad to lead up to it. *Sylvie and Bruno*, we learn from his idiosyncratic preface, was also written in a reverse manner. Instead of inventing a plot, then fitting scraps of his experience into it, he did just the opposite. He took odds and ends of his life—remembered incidents, remarks of children, old letters, half-forgotten poems he had written, even two ideas that came to him in dreams—then over many years worked out a plot that would weave together this "huge unwieldy mass of litterature" (the extra "t" was of course intended).

In keeping with his fondness for puzzles, Carroll posed two problems in his preface: to identify spots where he "padded" his narrative to bind fragments together, and to determine which stanzas of the mad Gardner's song were adapted to the plot and which suggested aspects of the plot. Both questions were answered in the preface to *Sylvie and Bruno Concluded*, where Carroll introduced more puzzles about his novel, promising to answer them in a book of original puzzles and games that he never got around to writing. As Florence Becker Lennon says in her *Life of Lewis Carroll*, the main unity behind his patchwork story is Carroll himself, with his unique blend of interests and beliefs.

Sylvie and Bruno had its origin in Carroll's short story "Bruno's Revenge," published in 1867 in *Aunt Judy's Magazine*. (With many changes, the story became chapters 14 and 15 of this book.) For a while his working title was "Four Seasons," presumably because the plot covers one year. The first volume, *Sylvie and Bruno*, was published by Macmillan in England a few days before Christmas 1889 and by the same company in the United States the following year. That Carroll thought of the book as partly nonfiction is indicated by his playful index. Volume 2, *Sylvie and Bruno Concluded*, issued shortly after Christmas in 1893 by Macmillan (in the United States in 1894), had a combined index for both volumes.

Although Carroll occasionally addressed his readers as children (for instance, "dear Child who reads this!" in chapter 14), he realized that much of his book could be understood only by adults. "It is a book intended for *all* ages—not for children only, as *Alice* was," he said in a letter. Two attempts were made by others to omit the adult episodes and limit the novel to scenes involving only Sylvie and Bruno. Carroll's youngest brother, Edwin, edited the first abridgement, *The Story of Sylvie and Bruno* (Macmillan, 1904). Philip Blackburn and Lionel White made the second abridgement for their anthology *Logical Nonsense: The Works of Lewis Carroll* (Putnam, 1934). The story has twice been dramatized for children, by Miss H. B. Griffiths in 1896 and by Roger Lancelyn Green in 1945, though neither play has been printed or acted.

John Ruskin once asked Carroll to write a novel that would not be a dream, but whether *Sylvie and Bruno* fulfilled this hope is debatable. Although the elderly narrator frequently falls asleep, Carroll makes clear that he is not dreaming in the conventional way. In the preface to the second volume, he distinguishes between two "psychical states": an "eerie" state in which the narrator is conscious of actual surroundings, but also conscious of fairies, and a "trance state" in which his soul "migrates to other scenes, in the actual world or in Fairyland." Using today's parlance, the second state is an OBE or "out-of-body experience." Such experiences were stressed by the theosophists of Carroll's time; indeed, he notes in his preface that the trance state is similar to "such as we meet with in 'Esoteric Buddhism,'" the title of a book by theosophist A. P. Sinnett.

The narrator, though never named, is obviously Carroll. At age fifty-five, when he was working on the novel, he speaks in a letter of "passing into the sere and yellow leaf" of his dotage, and throughout his adult life he was painfully aware of the age gap between himself and his child-friends. In *Sylvie and Bruno* he imagines himself to be seventy, a weary, lonely, bespectacled old man who periodically dozes on trains or when gazing into a fire. In the story "Bruno's Revenge" the narrator actually says to Bruno, "My name's Lewis Carroll," but in *Sylvie and Bruno* he keeps himself vague and anonymous. We are not told his profession, though we do learn that he knows a lot about mathematics and likes to sketch. He suffers from a heart ailment, perhaps symbolic of Carroll's sadness over the loss of child-friends, like Isa Bowman, who grow up, marry, and stop writing to him. In a letter to a child-friend just married, written between the publication dates of the two volumes of *Sylvie and Bruno*, he calls himself "a solitary, broken-hearted hopeless old bachelor." Harry Furniss, who illustrated *Sylvie and Bruno*, was instructed not to show the narrator in any of the pictures.

The book has three main settings: England, Fairyland, and Outland.

The nameless narrator is a Londoner. His doctor friend, Arthur Forester, lives in Elveston (note the "elves"), an imaginary town said to be modeled on Hatfield, where Carroll often stayed. It is reached by train, by way of Fayfield (note the "fay") Junction.

Fairyland, of which Elfland and Dogland are provinces, is to a large extent modeled on the fairyland of Shakespeare's *Midsummer Night's Dream*. Bruno, we are told, had been a servant of King Oberon, and Queen Titania is also mentioned (chapter 24).

Outland is an outlandish parody of Oxford, with its wardens, subwardens, and learned professors. As we learn in chapter 6, it is a thousand miles from Elfland. Dogland, some fifty miles from Outland, has its own language, Doggee, which all fairies understand. Raymond Queneau, a noted French novelist, wrote an entire article (translated in *Jabberwocky*, Winter 1977) on Doggee syntax. *Wow* means "What do you want?" *Hooyah wah* is "Come in." There is even a word for "not to be"—*wooh*. In the second volume we meet the dog king again and learn a new Doggee word, *bosh*. It means the same thing as in English, and is pronounced with half a cough and half a bark.

Sylvie (from the Latin name *Sylvia*, "female forest dweller") is nine or ten, with long curly brown tresses and big brown eyes. Like her brother Bruno, after she becomes a fairy her normal height is only a few inches, small enough to sit on a mushroom and produce "teeny-tiny" music by stroking a daisy's petals. Her main task seems to be taking care of Bruno—making him study his lessons, correcting his speech and behavior, and trying to keep him out of mischief. She is, of course, Carroll's ideal child—beautiful, loving, innocent, pure in heart. At the close of the second book she becomes an angel.

Bruno (from the Old High German word for "brown") is about five, with brown hair and eyes like his sister. He is extremely clever, as merry, mischievous, and hyperactive as Robin Goodfellow (alias Puck), and smart enough to know something about Shakespeare, as we learn from the amateur theatrical he performs (chapter 24) for an audience of frogs. I suppose it's a coincidence, but change "Robin" to "Robun," and the name is an anagram of "Bruno."

Bruno's adult knowledge contrasts sharply with his babyishness. He frequently sucks his thumb, and speaks in a baby talk that Carroll thought he based on the actual talk of children, but surely no English child ever talked like Bruno. Although baby talk was a convention in Victorian fiction, Bruno's "oo's" and "welly's" must have been almost as hard to take by Carroll's readers as they are today. Evelyn Waugh (reviewing *The Complete Works of Lewis Carroll* in the *Spectator*, October 13, 1939) calls Bruno "a creation of unique horror, who babbles throughout in baby-talk, like the 'control' of a 'medium.'" Happily, he drops his infantile accents when he sings. *Sylvie and Bruno*, many critics have noted, is the only work of Carroll's in which a boy is likable.

Critics have also observed that Bruno is a common name for both bears and dogs; also that it begins with the famous "B" of Carroll's *Snark*. Cats dominate the *Alice* books, but in this book the dog is dominant. Did Carroll associate cats with girls, dogs with boys?

More unbearable than Bruno's baby talk is the incessant hugging and kissing that goes on between the two siblings. It is, of course, as innocent as Carroll's constant kissing of little girls, but it occurs so often that it is as hard to avoid speculating about unconscious sexual emotion as it is to avoid similar speculation about such emotions in Carroll's fondness for photographing attractive little girls in the nude.

His Imperical Fatness, Prince Uggug (note the "ugh" and "ugly" in his name), is everything Carroll hated in boys. He is the same age as his cousin Sylvie, but stupid, fat, and cruel. In *Alice in Wonderland* the Ugly

Duchess's baby boy turns into a pig. In *Sylvie and Bruno Concluded*, Uggug's "prickly" nature turns him into a porcupine.

The benevolent Warden of Outland, father of Sylvie and Bruno, is for a time disguised as a beggar. Eventually he becomes King of Elfland. His evil younger brother Sibimet is the Sub-Warden and father of Uggug. In the preface to the second volume Carroll writes: "May I take this opportunity of calling attention to what I flatter myself was a successful piece of name-coining. . . . Does not the name 'Sibimet' fairly embody the character of the Sub-Warden? The gentle Reader has no doubt observed what a singularly useless article in a house a brazen trumpet is, if you simply leave it lying about, and never blow it!"

The remark about the brazen trumpet is clear: Carroll is defending his right to blow his own horn. But what in the world does "sibimet" mean? After some inquiries, the mystery was finally resolved by two correspondents of mine: Denis Crutch in England (whose interpretation came by way of Dr. Selwyn Goodacre) and Everett Bleiler, an erudite friend in New Jersey. *Sibi* is a Latin third person reflexive pronoun, in the dative case, with its use as a name in this instance implying a person concerned mainly with himself. The suffix *–met* intensifies the pronoun. A rough English equivalent would be "his very self," with implications of selfishness. Ruth Berman suggests that Carroll may have also had in mind the word "sibilant," suggesting the sound of a snake. (My wife noticed that the first two consonants of Sibimet are "S" and "B," the initials of Sylvie and Bruno, but this may be sheer coincidence.) I assume that Tabikat, the name of Sibimet's evil wife, is a play on "tabby cat," but because *tabes* is Latin for disease and moral decay, Carroll may have intended a portmanteau word meaning a morally depraved cat.

Roger Lancelyn Green believes, and Morton Cohen concurs, that Sibimet and Tabikat are caricatures of Alice Liddell's father and mother. Carroll had little admiration for either. After all, it was Mrs. Liddell who stopped Alice from seeing him and who burned all his letters to Alice. Mrs. Liddell had a reputation for being "catty." Professor Cohen called my attention to Tabby's remark in chapter 5, "And am I Vice-Wardeness?" and reminded me of the anonymously written epigram that made its rounds at Oxford:

I am the Dean and this is Mrs. Liddell.
She plays the first, and I the second fiddle.
She is the Broad; I am the High;
And we are the University.

Sibimet's attitude toward his sons is one of constant disrespect. In one episode he hangs his hat on Uggug as if he were no more than a hatrack. In another, he beats him over the head with an umbrella as if driving a loose nail into the floor. In her biography of Carroll, Anne Clark calls attention to the prototypes of Sibimet and Tabikat in a story Carroll wrote in his youth for his *Rectory Magazine*.

The Professor, a caricature of an absent-minded Oxford scholar, is (as we learn in the second book) probably an extraterrestrial, perhaps from the Moon. His colleague, the Other Professor, keeps his nose buried in books when he's not tripping over furniture.

The mad Gardener, whose song is the best nonsense poem of the story, could have strayed from Wonderland. He does frantic jigs while he sings, likes to stand on one leg when silent, and likes to water plants with an empty can because it's lighter to hold. Furniss has drawn him with wisps of straw sticking out from his hair and clothes. It was a Victorian convention to depict lunatics this way. You'll see similar straw in John Tenniel's drawings of the mad March Hare.

Arthur Forester (note the "forest") is an adult version of Bruno—handsome, wise, good, and devout. In a letter to Furniss, Carroll says he is "forty at least," and suggests he be drawn with a likeness to King Arthur. The doctor's whimsical ideas show that he has as much in common with the narrator as he has with Bruno.

Lady Muriel Orme (*orme* is French for "elm") is a grown-up Sylvie. Her age is about twenty-one, almost exactly as much younger than Arthur as Alice Liddell was younger than Carroll. She is as witty as Arthur and the narrator. Her father, the Earl of Ainslee, is said to be based on Carroll's friend Lord Salisbury of Hatfield.

Captain Eric Lindon (did Carroll have in mind a similarity to "linden," the name of a tree?)[3] is only a few years older than his cousin Muriel. Carroll portrays him as a religious skeptic who nevertheless is brave and good. In chapter 22 Lindon rescues Bruno from an onrushing express train, and in the second volume he saves the life of his rival

Arthur. Carroll transforms him into a theist at the end of the second book, intimating that he is on his way to becoming a Christian.

There is lots of magic in the story. Bruno is capable of making what he calls "phlizzes"—objects and even people that he conjures "out of air." In chapter 29 he materializes an entire nursemaid, using ventriloquism to make her talk! In the same chapter he creates a nosegay of rare flowers from India. We are told in volume 2 that he performs "conjuring tricks," presumably of the sort ordinary magicians do.

Sylvie's Magic Locket, given to her by her father, is the novel's central symbol. We are told originally that there are two lockets, both heart-shaped, one red, one blue. When Sylvie chooses the red one, with the inscription "Sylvie will love all," rather than the blue one, with "All will love Sylvie," her father weeps with joy. (Characters weep as often in this novel as Sylvie and Bruno hug and kiss.) In the concluding book we learn that the locket is really a single one that can be viewed two ways. "When you look *at* it, it's red and fierce like the sun—and, when you look *through* it, it's gentle and blue like the sky!" It has marvelous powers. In chapter 8 Sylvie uses it to make trees, and to turn a mouse into a lion on which Bruno rides.

The Professor's crazy inventions (and the inventions of his real-life counterpart, the German Professor Mein Herr in the sequel) provide the story's most delicious nonsense. We learn here about the Professor's boots with umbrellas to ward off horizontal rain, his portable plunge bath, his machine for shortening and stretching animals, and above all his Outlandish Watch. The episodes created by the watch are the second-earliest known instances in fiction of time travel made possible by a machine. (H. G. Wells's *The Time Machine* had appeared a year earlier in a magazine.) Unlike so many later science fiction writers, Carroll was well aware that travel into the past could generate logical contradictions. Going back in time, the narrator prevents a tragic accident, but when he returns to the present, the accident happens again just the way it did before. You can enter the past and think you are altering it, but the changes are unreal. The backward events that occur when the watch's Reversal Peg is turned are the first scenes in fiction in which time goes the wrong way. Note that although the events run backward, like a reversed motion-picture film, each sentence of the dialogue remains forward—Carroll's concession to readability. It is much harder to read backward spelling than to read a mirror-reversed poem like "Jabberwocky" by holding it up to a mirror.

Arthur doesn't build devices, but his "thought experiments" are as fantastic as the Professor's mad inventions. His elevator ideas in chapter 8 are striking anticipations of a famous thought experiment used by Einstein to explain the equivalence of gravity and inertia. Zero gravity prevails in a free-falling elevator, and objects inside would rise as though attracted by gravity from above if the elevator moved downward at a faster rate of acceleration than the objects inside. Later in the book Arthur suggests that the reason we don't see the images on our retinas as inverted (which they are) is because our brain is also upside down. A Sillygism, he explains in an often quoted passage, consists of combining two prim Misses to produce a Delusion.

We have already mentioned the Gardener's song as one of Carroll's finest nonsense poems. Its great last stanza, in *Sylvie and Bruno Concluded*, is:

He thought he saw an Argument
That proved he was the Pope:
He looked again, and found it was
A bar of Mottled Soap.
"A fact so dread," he faintly said,
"Extinguishes all hope!"

After finishing, the Gardener chokes up with sobs. "That song is his own history, you know," explains the Professor. Tears glitter in Bruno's eyes when he remarks, "I's *welly* sorry he isn't the Pope! Aren't *you* sorry, Sylvie?"

"Well—I hardly know," his sister answers. "Would it make him any happier?"

"It wouldn't make the *Pope* any happier," says the Professor.

All this banter is Carroll at his Carrollian best. In contrast, consider the long ballad recited by the Other Professor in chapter 11 that tells about a cruel April Fools' joke. Exploiting the ancient phrase about robbing Peter to pay Paul, a phrase originally applied to St. Peter and St. Paul, the ballad has nothing to do with the story. It struck me as one of Carroll's longest, most boring poems until I came across Roger Lancelyn Green's note in his edition of Carroll's *Diaries* (New York, 1954, p. 404) stating that the ballad is a parody of a famous 1785 speech to Parliament by Edmund Burke. Titled "On a Motion Made for the Papers Relative to the Nabob of Arcot's Debts," the speech was a savage indictment of England's cruel, avaricious policy toward India.

Arcot was a district in India's Carnatic region. The East India Company had installed a Muslim prince as the region's nabob (governor) and furnished him mercenaries for expanding his domain. But the nabob proved to be so inept that he allowed the Carnatic region to be devastated and plundered by an invading army. The British moved in troops to take over the region, but instead of providing funds to feed the starving, and helping the region recover, they imposed a heavy burden of debt that the impoverished region could not pay. When the debt reached about two million pounds, the British "generously" provided more loans at usurious rates that soon doubled the debt to four million. The debts were in the form of fraudulent mortgages provided by unscrupulous private moneylenders of whom one Paul Benfield was chief. In Carroll's poem, Peter is India, kept "sweetly picturesque in rage" by the cruel "loans" of Paul, who represents England's moneylenders. In Burke's strong words, the nabob of Arcot's debts were a "foul, putrid mucus" imposed by "tapeworms, which devour the nutriment, and eat up the bowels of India."[4]

Note that Alice's birthday of May 4 appears in the poem's second stanza. Peter, tearing out his yellow curls and later acquiring a wig, calls to mind the youthful yellow ringlets of the wasp in a wig, in the famous lost episode of *Through the Looking-Glass*.

Like the *Alice* books, *Sylvie and Bruno* is laced with wordplay. There are hundreds of puns, most of them not memorable, and the narrator has an annoying habit of explaining lots of them by making comments or italicizing the relevant words. The constant quibbling by Bruno and others is much funnier. Like so many characters in Wonderland and behind the mirror, Bruno is fond of taking statements so literally that their meanings become absurd. Uggug is too fat to go through a hole in a target, Bruno remarks, after being told that Uggug "went in right *here*." When Tabikat asks her husband what he got a dagger for, he says he got it for eightpence. As in *Alice*, there are numerous instances of amusing logical contradictions. The distance between Outland and Fairyland is five times longer in one direction than the other. A crocodile crawls along its tail, up its back, and down its own head to its nose. The Professor tells Sylvie that the Other Professor, on three occasions, started reciting a poem and never stopped.

When Carroll chose Harry Furniss to illustrate his novel, Tenniel warned Furniss that he would soon be in serious trouble with Carroll over the pictures. The warning was accurate. Their relationship became so strained that at one point Furniss refused to continue. Carroll was constantly sending him sketches to show how he wanted his characters to look (fifteen are reproduced by Edward Guiliano in a book he edited, *Lewis Carroll Observed*), recommending various little girls as model and debating with Furniss about how Sylvie and Bruno should be clothed. In one letter Carroll wrote: "I *wish* I dared dispense with *all* costume; naked children are so perfectly pure and lovely, but Mrs. Grundy would be furious—it would never do." Carroll proposed a transparent dress for Sylvie, but Furniss decided it would be best to make it opaque, though filmy and clinging. No wings on the pair and no high heels on Sylvie were two of Carroll's firmest injunctions.

The novel's two plots, which interlock as the narrator shuttles back and forth between England and Outland or Fairyland, are so hard to follow on a first reading that perhaps brief summaries (including the action of *Sylvie and Bruno Concluded*) will be helpful.

Like *Finnegans Wake*, the story opens in the middle of a sentence. I take this to indicate that the narrator has abruptly experienced his first OBE. He finds himself in Outland, invisible, though capable of seeing and hearing everything. The Warden of Outland is about to leave for Elfland. His crafty brother Sibimet, the Sub-Warden, tricks him into signing a document that makes the brother and his simpleminded wife the equivalent of king and queen, and their disgusting son Uggug heir to the throne.

The Warden's children, Sylvie and Bruno, aided by the court Professor, run off to Elfland, with a stopover in Dogland. Their father is made King of Elfland, and Sylvie and Bruno become tiny fairies. The Professor later takes them back to Outland to help block the Sub-Warden's power grab. To delay the Coronation Banquet, the Professor begins a lengthy lecture. It is interrupted by a brief, mysterious hurricane which I believe Carroll intended to be a whirlwind from heaven, like the wind that rescued and transformed Coleridge's Ancient Mariner. When the King of Elfland returns, his brother, cleansed by God's whirlwind, instantly repents. The King of Elfland forgives him and allows him to remain the ruler. Uggug turns into a porcupine. Everybody except Uggug is happy. As the sun sets, Sylvie becomes an angel who whispers, from a "darling blue" sky, "It is love."

The other plot, taking place in the real world of England, begins with the narrator's journey by train to Elveston to see his old friend Dr. Arthur Forester. Arthur, a country doctor, has long been in love with Lady Muriel, but because he is too poor to marry, and perhaps too old for Muriel, he has concealed his feeling. Lady Muriel loves Arthur, but not knowing how he feels about her, and supposing herself bound to her cousin Captain Eric Lindon, she also keeps her distance from Arthur. Muriel has mixed emotions about Eric. She is fond of him, though dismayed by his unshakable atheism.

Arthur inherits a fortune, but believing Lady Muriel to be in love with Eric, he still does not tell her how he feels. After an army promotion, Eric proposes to Muriel and she accepts. Heartbroken, Arthur heads for India.

When Lady Muriel tells Eric about her fears that she would be unhappy married to a non-Christian, he gallantly releases her from their engagement. Arthur returns to Elveston to propose to Muriel. No sooner are they betrothed than Arthur is compelled to leave to help the residents of a nearby town battle an epidemic. After a false report of his death, Eric finds him and brings him home in a crippled, debilitated state. Presumably Muriel will nurse him back to health and they will live happily ever after.

Compared with the flood of literature about *Alice*, surprisingly little has been written about *Sylvie and Bruno*, though in recent years critical interest in the novel may have been slowly rising. Carroll was not, as he himself realized, a skillful writer of fiction. The consensus on *Sylvie and Bruno* is that, aside from its wordplay and nonsense, it is a flawed, embarrassing work, produced at a time when Carroll correctly perceived that his creative powers were waning. It is difficult for modern readers to be impressed by the book's pious exhortations. In the first volume's preface Carroll sounds exactly like one of today's television evangelists when he urges his readers to get right with God because at any moment they may die. His rule for those who attend the theater, as he himself loved to do, is to ask whether they would like to be enjoying the play if death suddenly struck. "The safest rule is that we should not dare to *live* in any scene in which we dare not *die*." Carroll was so incensed by sex on stage and by what he considered blasphemy that he wanted to Bowdlerize Bowdler. A new edition of Shakespeare should be written for girls, he maintained, because Bowdler's version was insufficiently expurgated!

Death and the mystery of evil pervade the novel, from such playful scenes as Bruno using a dead mouse for a ruler or a pillow, to one in which Sylvie lies on the grass, sobbing uncontrollably over a dead rabbit. "And God meant your life to be so beautiful!" she cries. One need not be a Christian to take an interest in Arthur's opinions about good and evil, free will, prayer, and other theological topics, but readers with no faith in a personal God and an afterlife will be singularly unimpressed.

In *Sylvie and Bruno Concluded* Arthur strongly opposes the notion that eternity in heaven will be dull. On the contrary, there will be an infinity of new things to be learned, even in mathematics, and new experiences to undergo. His greatest departure from conservative Anglican theology is an impassioned conjecture that because of the enormous influence of bad environment on behavior there will be far fewer lost souls than we imagine. Carroll, by the way, did not believe that the Christian hell meant an eternity of suffering. He even published a pamphlet on the topic and said that if he believed Jesus had taught such a doctrine he would stop calling himself a Christian!

On the deepest metaphorical level Carroll surely intended the contrast between England and Fairyland to reflect a Platonic dichotomy between the world we know and the transcendent world of God. Like the narrator, we lead double lives. Our bodies are trapped for a brief time in the fallen world, the lost Paradise, but our true home is the Paradise we will enter when we have our last and greatest OBE. Arthur's cry of "Look Eastward!" as he views a rising sun on the last pages of this book, suggests the Easter hope of the Resurrection. Compared to the new dimensions we are destined to enter, our life here among the shadows on the wall of Plato's cave is indeed a kind of dream. The *Alice* books end with Alice wondering if she dreamed her adventures or if she herself is no more than a phantom in the mind of the dreaming Red King. As Ruth Berman observes in her valuable essay "Patterns and Unification in

Sylvie and Bruno," each Alice volume starts and ends in the real world, but *Sylvie and Bruno* starts in Outland, and in the sequel ends in either Outland or Fairyland.

How should we finally assess this long, complicated, curious "novel of ideas," so rich in nonsense, linguistic play, and philosophical reflections? Victorian popular fiction often combined religious piety with sentimentality—the prolonged death of Dicken's Little Nell is a frequently cited example—but there is no use denying that *Sylvie and Bruno*, more than most, oozes with sentimentalism, mawkish moralizing, and a cloying sweetness "as unpalatable," in Miss Berman's words, "as too much cotton candy." Miss Lennon called it Carroll's "biggest doodle," a "dreadful book," a book of "infernal dullness." Yet she could also exclaim, "What a book! What a noble ruin!"

Derek Hudson, in his *Lewis Carroll: An Illustrated Biography* (1977), said it best. *Sylvie and Bruno* is "one of the most interesting failures in English literature. It is certainly unique; no one but Dodgson could have written it; nothing like it will be produced again."

NOTES

1. For this observation of how cleverly Carroll attached *Sylvie and Bruno* to the second *Alice* book I am indebted to Brian Sibley's informative essay, "The Poems in *Sylvie and Bruno*," in *Jabberwocky* (the periodical of England's Lewis Carroll Society) (Summer 1975): 51–58. It is the first full discussion of all the poems in the novel. The same issue also features a splendid article by Denis Crutch, "*Sylvie and Bruno*: An Introduction," and a detailed "Bibliography of *Sylvie and Bruno*," by Selwyn Goodacre.

2. The full text of this letter, written in 1892 to the Reverend Charles Alfred Goodhart, can be found in *The Letters of Lewis Carroll*, edited by Morton N. Cohen and Roger Lancelyn Green (New York: Oxford University Press, 1979), 2: 885.

3. The suggestion that Eric Lindon and Muriel Orme have last names based on trees was made by Alfreda Blanchard in her entertaining article "Sylvie and Bruno Re-opened," *Jabberwocky* (Winter 1983/84): 14–17.

4. Burke's complete speech is given in *The Speeches of the Right Honourable Edmund Burke* (London, 1816). An abridged version appears in *Irish Orators*, edited by Thomas Kettle (London, 1915). I wish to thank Ruth Berman and Russell Barnhart for their help in obtaining copies of Burke's speech.

33.

PHANTASMAGORIA

This is my introduction to the Prometheus Books edition (1998) of Lewis Carroll's second-most memorable poem. The first, of course, is his great nonsense ballad *The Hunting of the Snark.* I annotated it for Simon & Schuster in 1962, and a handsome much enlarged edition was published by W. W. Norton in 2006.

It has been said many times that the serious poetry of Lewis Carroll (the pen name of Charles Lutwidge Dodgson) tended to be humdrum, but that his comic verse was far superior. Short specimens of humorous verse are scattered through his two *Alice* books, and his two *Sylvie and Bruno* books. *The Hunting of the Snark*, a long narrative poem in what Carroll called eight fits, is, of course, his masterpiece of nonsense verse. *Phantasmagoria*, in seven cantos, is his less well-known comic narrative.

Unlike *The Snark*, *Phantasmagoria* is not nonsense. It tells an entertaining tale about a middle-aged man of forty-two who is haunted by a friendly but inexperienced ghost. Its 150 stanzas are artfully constructed, with impeccable rhythms and clever, at times highly unusual, rhymes. It is a very funny poem.

Although *Phantasmagoria* does not take ghosts and goblins seriously, Carroll had a lifelong interest in the paranormal. He joined the Society for Psychical Research shortly after it was formed in 1882, and his library contained many books on spiritualism and other occult topics. As a devout Anglican, he firmly believed in life after death (though he denied the doctrine of eternal suffering in hell). And while he believed that physical phenomena produced by mediums were genuine, he did not think they were the work of departed souls. A letter that Carroll wrote a friend, Langton Clarke, about this is often quoted in part in biographies of him. Here is the letter in full as it appears in Morton Cohen's *The Letters of Lewis Carroll*:

> My dear Langton Clarke,
>
> I keep making efforts to write off my arrears in letter-writing—but with new letters needed every day, it seems a hopeless task: however, one of the oldest of them shall go today—*viz.*, your letter of March 14, about a chemical explanation of the "two rings" in Zöllner's marvelous book.* I think the explanation so good as to make it very highly probable that the thing was trickery, in that case: but that trickery will *not* do as a complete explanation of all the phenomena of table-rapping, thought-reading, etc., I am more and more convinced. At the same time, I see no need as yet for believing that disembodied spirits have anything to do with it. I have just read a small pamphlet, the first report of the Psychical Society, on "thought-reading." The evidence, which seems to have been most carefully taken, excludes the possibility that "unconscious guidance by pressure" (Carpenter's explanation) will account for all the phenomena. All seems to point to the existence of a natural force, allied to electricity and nerve-force, by which brain can act on brain. I think we are close on the day when this shall be classed among the known natural forces, and its laws tabulated, and when the scientific skeptics, who always shut their eyes, till the last moment, to any evidence that seems to point beyond materialism, will have to accept it as proved fact in nature. You would find the "Report" (published by Trubner for 2*s*.) very interesting—all the more so that "thought reading" is a phenomenon on which any domestic circle can experiment for themselves: it needs no professional "medium."
>
> With love to Margie & Co., I am
> Sincerely yours,
> C. L. Dodgson

---

*There is nothing "marvelous" about this book. Johann Carl Friedrich Zöllner was an astronomer at the University of Leipzig. A gullible, stupid man, totally ignorant of conjuring methods, he fell completely for the crude trickery of the American medium Henry Slade. Zöllner's book *Transcendental Physics* (1878), extolling Slade's great powers to move objects in and out of the fourth dimension, was partly translated into English in 1880. For an account of this American mountebank, see my article on Slade in *The Encyclopedia of the Paranormal*, edited by Gordon Stein (Amherst, NY: Prometheus Books, 1996).

Although Carroll believed in what today is called ESP and PK (psychokinesis), he did not think that one could communicate with the dead, or that houses could be haunted by mischievous spooks. He would have followed with interest the later history of parapsychology, though he surely would have been appalled by the gullibility of Conan Doyle, who actually was convinced that real fairies frolicked in English gardens.

Phantasmagoria and Other Poems was published in 1869. It contained thirteen comic poems followed by thirteen serious ones. The book had no illustrations. The title poem was published a second time in 1883 in an anthology titled *Rhyme? and Reason?* a book profusely illustrated with sixty-five sketches by Arthur Burdett Frost (1851–1928). Carroll had been unsuccessful in finding a British artist willing to illustrate the book before he finally persuaded Frost to take it on.

Frost was one of the most popular American illustrators of his time and the first American artist to become well known in England. In addition to prolific work for such American magazines as *Harper's*, *Scribner's*, *Colliers*, and *St. Nicholas*, he also contributed to British periodicals, illustrated books by Dickens and Mark Twain, but is best remembered today for his drawings in six Uncle Remus books by Joel Chandler Harris. He also authored books of his own, with such titles as *The Bull-Calf and Other Tales*, *The Golfer's Alphabet*, *Stuff and Nonsense*, *Sports and Games in the Open*, and *Carlo*. Born in Philadelphia, Frost visited England in 1877–1878 where he and Carroll met, lived for a few years in Paris, and died in Pasadena, California, at the home of his son John, a landscape painter. He always insisted that his art was entirely self-taught.

The eleventh edition of *The Encyclopedia Britannica*, in its article "Caricature," had this to say about Frost:

> Entirely native in every way is the art of A. B. Frost (b. 1851), a prominent humorist who deals with the life of the common people. His caricature (he is also an illustrator of versatility and importance) is distinguished by its anatomical knowledge, or, rather anatomical imagination. Violent as the action of his figures frequently is, it is always convincing. Such triumphs as the tragedy of the kind-hearted man and the ungrateful bull-calf; the spinster's cat that ate rat poison, and many others, force the most serious laughter by the amazing velocity of action and their

> unctuousness of expression. Frost is to American caricature what "Artemus Ward" has been to American humor, and his field of publication has been chiefly the monthly magazine.

Carroll's extensive correspondence with Frost was filled with high praise for most of his preliminary sketches for *Rhyme? and Reason?* and later for Carroll's *A Tangled Tale*. On the other hand, Carroll consistently raised objections to certain drawings, and offered suggestions for redoing them. It was a practice that infuriated other illustrators for Carroll's books, and one that led to an eventual cooling of his relationship with Frost. Several of Carroll's letters to Frost can be found in Cohen's *The Letters of Lewis Carroll*. Here is his first letter, dated January 7, 1878:

> Dear Sir,
>
> Excuse the liberty I am taking in addressing you, though a stranger. My motive for doing so is that I saw a page of pictures, drawn by you in *Judy* last month, on "The Eastern Question" as discussed by 2 barbers, which seemed to me to have more comic powers in them than anything I have met with for a long time, as well as an amount of good drawing in them that made me feel tolerably confident that you could draw on wood for book illustrations with almost any required amount of finish.
>
> Let me introduce myself as the writer of a little book (*Alice's Adventures in Wonderland*) which was illustrated by Tenniel, who (I am sorry to say) will not now undertake woodcuts, in order to explain my enquiry whether you would be willing to draw me a few pictures for one or two short poems (comic) and on what sort of terms, supposing the pictures to range from 5x3-1/2 downwards to about half that size, and to have about the same amount of finish as Tenniel's drawings usually have. Believe me
>
> Faithfully yours,
> C. L. Dodgson

"It is difficult to find words which will express, as strongly as I wish," Carroll said in opening a letter to Frost in 1881, "how *thoroughly* I admire your pictures to the ghost-poem. They really are *wonderful*." Nevertheless, Carroll returns to Frost a sketch of the poem's narrator, saying he does not like the way the man is portrayed. He recommends that his nightshirt be replaced by a flowing dressing gown. Instead of holding a warming pan with which to hit the ghost, Carroll thinks a pillow would be "more hopelessly useless for exterminating ghosts, and therefore more comic." He hopes Frost will be able to paste a new drawing over the man rather than have to redo the entire picture.

Figure 29: The poem's narrator threatening to hit the ghost with a pillow.

As you will see, Frost followed these suggestions, as well as Carroll's many proposals for redrawing other illustrations in *Rhyme? and Reason?*

I hope my notes to the text will add interest to a poem long neglected by today's Carrollians.

34.

THE NURSERY ALICE

This is my introduction to the Dover paperback edition (1966) of Carroll's *Nursery Alice*.

"It is one of the mysteries of publishing that this charming book . . . has been out of print for so many years. As a book for children under five, it is only surpassed by the best of Beatrix Potter. . . ." So wrote Roger Lancelyn Green, commenting on *The Nursery Alice*, in *The Diaries of Lewis Carroll*, which he so admirably edited.

Since Green made those comments, in 1954, there have been rumors of forthcoming reprints of *The Nursery Alice*, by both English and American publishers, but the promised books failed to materialize. Now, as a companion piece to its earlier facsimile of *Alice's Adventures Under Ground*, the publisher has produced this facsimile volume of the second edition of *The Nursery Alice*. It would have been easy to reproduce from the first, but since Carroll thought the colors of that edition too gaudy, it seemed best to use the one he approved.

The first mention of *The Nursery Alice* in Carroll's *Diary* is an entry on March 29, 1885. "Never before," he writes, "have I had so many literary projects on hand at once. For curiosity I will here make a list of them." Fifteen projects are cited. The ninth is *The Nursery Alice*, "for which twenty pictures are now being coloured by Mr. Tenniel." On July 10 he records that Tenniel had finished the pictures, but it is not until December 28, 1888, almost four years later, that he writes: "Began text of *The Nursery Alice*." Less than two months later, on February 20, 1889, an entry reads: "Sent off last of MS. for *The Nursery Alice*." On April 18 he writes that he completed checking page proofs.

But Carroll was a difficult man to please with respect to a book's appearance. Charles Morgan, in *The House of Macmillan: 1843–1943* (Macmillan, 1943), says that ten thousand copies of *The Nursery Alice* were printed, but that Carroll took one look at them and decided that the pictures were too gaudy. "No copy, he said,"—I quote from Morgan—"was to be sold in England; all were to be offered in America. They were offered, and declined as not being gaudy enough."

A second edition of ten thousand was scheduled, but now Macmillan made sure that Carroll saw advance copies, completely bound, before the final print run. Carroll rejected the uncolored samples sent to him in September 1889 because the March Hare on the back cover was off-center. "As to how many copies we can sell I care absolutely nothing," he wrote to his publishers (I quote from Green's *Lewis Carroll*, a Bodley Head monograph published in 1960); "the only thing I *do* care for is, that all copies that *are* sold shall be artistically first-rate." A dozen copies with color, sent to him in October, were also returned, because the covers cracked when he opened them and the leaves had a tendency to curl.

Samples were finally approved by the end of October, and pages for the new edition were printed by late February 1890. "Received from Mr. E. Evans a finished set of the sheets of *The Nursery Alice*," Carroll wrote in his *Diary* on March 7, 1890. "It is a *great* success. We can now publish at Easter." On March 25, Lewis Carroll speaks of a trip to London to inscribe a hundred copies of what must have been bound volumes.

Exactly what happened to all ten thousand copies of the first edition is not known. There is a record of four thousand having been sent to the United States, and five hundred to Australia. Many were probably given to hospitals in England and elsewhere. Copies of the first edition are said to be identifiable by the brighter colors of the pictures and by an off-white paper in contrast to the purer white of the second edition, but points on the variant states of the two editions are far from established, and the situation is one of great bibliographic confusion.

There is also a mystery about whether Tenniel himself actually colored the pictures. Carroll's *Diary* speaks as if he did, and in a letter written as late as April 1, 1889, he writes of hoping to get the book published by Easter, and that it contains "pictures enlarged and coloured by Tenniel." But when the book was advertised as "in preparation" in Macmillan's 1886 facsimile of *Alice's Adventures Under Ground*, its pictures are said to be "enlarged and coloured under the Artist's superintendence." In the Macmillan edition of this book, a similar advertisement, as well as its title page, speak of "coloured enlargements from Tenniel's illustrations." What probably happened was that Carroll originally hoped that Tenniel would do the job himself, but that Tenniel finally delegated the work to someone else. (Later, in 1911, *Alice's Adventures in Wonderland* and *Through the Looking-Glass* were brought out by Macmillan in one volume, with sixteen pictures that *were* colored by Tenniel.)

Carroll liked to dedicate his children's books to child-friends, often concealing their names in acrostic poems. *The Nursery Alice* is dedicated to Marie Van der Gucht, whose full name is read by taking the second letter of each line in the dedicatory poem. Marie was a friend of Climene Mary Holiday, a niece of Henry Holiday, the man who illustrated *The Hunting of the Snark*. Carroll's *Game of Logic*, included in *Symbolic Logic and the Game of Logic*, had earlier been dedicated to Climene with a similar second-letter acrostic.

Carroll first met Marie on July 24, 1885, when she was eleven. His *Diary* records taking her to see *The Mikado* on April 10, 1886, and traveling again to London on September 1 to escort her to the beach at Eastbourne. "Marie and I," he writes the following day, "after a little Bible-reading and letter-writing, spent the morning on the beach. . . ." On November 15 he obtained Marie's mother's permission for the girl to pose as Sylvie in Harry Furniss's illustrations for *Sylvie and Bruno*, but Furniss later revealed that he never took this suggestion seriously, and used his own daughter as the model. On December 30, 1886, Carroll took Marie and another girl to see the operetta of *Alice in Wonderland*, in London. Marie was sixteen in 1890 when *The Nursery Alice* was published. There are two more entries about her in the *Diary*. On August 24, 1895: "Dear Marie Van der Gucht came on a visit to me." And on August 30: "Marie went home again. Her week with me has been very pleasant, to both of us, I think. The Schusters have twice had her up to their house for lawn-tennis; and she and I have spent several evenings at Devonshire Park."

Why Carroll retold in simpler terms the story of Alice's first dream is plain enough from his sentimental preface, but one suspects that another reason may have been that he wanted a book of his own to give away to very young girls when he met them on trains or at the seaside. An entry in his *Dairy* (October 1, 1891) reads: "I made friends with a sweet-looking little girl, Constance Linnell, grand-child of the painter. Of course I afterwards sent her *The Nursery Alice*." The "of course" suggests that many copies of this book were given away to the "dimpled darlings" who were too young to read the other *Alice*. In Carroll's opinion, the age at which a child became too old for *The Nursery Alice* varied with the child's economic background. Florence Becker Lennon, in her *Life of Lewis Carroll*, quotes an amusing passage on this from a letter Carroll wrote to his friend Gertrude Thomson, the artist who illustrated his book of poems, *Three Sunsets*, and who designed and colored the covers of *The Nursery Alice*:

> I have just promised to give the little girl, of the porter who always carries my luggage, a book: and had intended it to be *The Nursery Alice*, as the child is 10, and I consider children of the lower orders to be 2 or 3 years behind the upper orders. But a lady, whom I consulted, advised me to give the real *Alice*, as probably more interesting, even now, to the child (they certainly do get very well taught now-a-days), and certainly of more permanent interest.

How successful is *The Nursery Alice* when read today to an English or American boy or girl, upper or lower class, age zero to five? I prefer not to guess. In some ways the language seems patronizing, but one must admit that Carroll has retold Alice's dream in a way that is easily understood by small children. The story has been shortened to about one-fourth its original length, the verse (except for the nursery rhyme about the Knave of Hearts) has been left out, and of course everything has been skillfully simplified. The only new episode is one about a puppy named Dash who doesn't like his oatmeal porridge (page 22), a digression so dull that one suspects Carroll must have put it in because it referred to an actual incident involving the pet of one of his child-friends. There is a delightful bit of wordplay (page 19) about the room being as full of Alice as a jar is full of jam. At two places he suggests something whimsical to do with the book itself: shaking it to make the White Rabbit tremble with fright (page 2), and bending up a lower corner of page 36 to see Alice (on the under leaf) looking up at the Cheshire Cat's grin (on the upper leaf).

This was the first time that color had been used for the *Alice* illustrations in book publication, and Carroll took advantage of the opportunity to add many descriptive color words that he had not used before: the green eyes of the Cheshire Cat, the Mad Hatter's yellow tie with red spots, the red flamingo in the croquet game, and so one. His comments on Tenniel's drawings show how seriously he took the illustrations. The Blue Caterpillar's nose and chin, Carroll explains (page 27), are really two legs. On page 53 he carefully identifies all twelve jurors in the picture. He calls attention to the foxglove plant on page 34, growing near the tree in which the Cheshire Cat is grinning, and explains that the foxglove was once called "Folk's-Gloves," with reference to the fairies who were spoken of as the "good folk" or "little folk."*

A note headed "Cautions to Readers," that Carroll added to the book's advertising pages (following the Easter and Christmas Greetings in the original edition), is worth a comment. (Its second part refers to an unkind review of *Alice's Adventures in Wonderland* that had been written by someone named Edward Salmon. Salmon had accused Carroll of borrowing from a book by Tom Hood that actually had been published nine years after the *Alice* book. Carroll's restrained reply does no more than correct Salmon's error on the date of Hood's book. (Lengthy quotations from Salmon's review can be found in appendix B of Mrs. Lennon's biography of Carroll.)

The real Alice, Alice Liddell, always remained in Carroll's memory as his favorite child-friend. She was in her late thirties and happily married when *The Nursery Alice*, the fourth and last of his published *Alice* books, appeared. Carroll was fifty-eight. His first meeting with Reginald Gervis Hargreaves (who had married Alice in 1880) occurred on November 1, 1888, two months before he began to write *The Nursery Alice*. "It was not easy," Carroll then wrote in his *Diary*, "to link in one's

*According to J. Worth Estes and Paul Dudley White (in their article, "William Withering and the Purple Foxglove," *Scientific American*, June 1965), an equally respectable theory makes the word a corruption of "foxes-glew," an old Saxon word for "fox music." This was a type of music produced by an ancient instrument consisting of bells hanging from an arched support, and which the foxglove blossoms presumably resemble as much as they resemble gloved fingers. The plant's leaves are a source of digitalis, a drug used in treating heart disease.

mind the new face with the olden memory—the stranger with the once-so-intimately known and loved 'Alice,' whom I shall always remember best as an entirely fascinating little seven-year-old maiden."

On December 9, 1891, a year after the appearance of *The Nursery Alice*, the last entry about Alice Liddell appears in Carroll's *Diary*. She had been visiting the Deanery of Christ Church and he had invited her to his rooms for tea. "She could not do this, but very kindly came over, with Rhoda (her sister), for a short time in the afternoon."

One has a feeling that, in the closing line of *The Nursery Alice*, Carroll was speaking directly to Alice Liddell; his words echoing that sad farewell of the White Knight as he rode slowly and precariously off into the chessboard forest behind the looking-glass. "Good-bye, Alice dear, good-bye!"

35.

ALICE'S ADVENTURES UNDER GROUND

The article that follows was my introduction to Dover's facsimile edition (1965) of Carroll's hand-lettered, self-illustrated manuscript.

It is not widely known that the Reverend Charles Lutwidge Dodgson, better known as Lewis Carroll, wrote three, possibly four, quite different versions of *Alice's Adventures in Wonderland.*

It all began on July 4, 1862. That was the "golden afternoon" on which Carroll and his friend Robinson Duckworth took three little girls on a boating trip up the Isis, a tributary of the Thames. The girls were the daughters of Henry George Liddell (his name rhymed with "fiddle"), dean of Christ Church, Oxford, where Carroll taught mathematics. Carroll was especially fond of Alice Liddell, then ten. It was mainly for her that he began his story of another Alice's tumble down the rabbit hole, inventing the whimsical details as he went along.

The "interminable fairy-tale," as Carroll referred to it in his diary, lasted through many later boating trips. Carroll assured Alice that he would write it all down for her, but it was not until February of the following year that the task was completed. This first manuscript, which he called *Alice's Adventures Under Ground*, was probably destroyed by Carroll in 1864 when he prepared a more elaborate hand-printed copy with thirty-seven pictures drawn by himself. We cannot be sure, but it seems likely that this second manuscript differed in many respects from the first. At any rate, we do know that on November 26, 1864, Carroll gave it to Alice as a Christmas present.

While Carroll was preparing this second version of his story, friends were urging him to find a publisher for it. He set about revising and expanding until the story was almost twice as long as it had been, and considerably more sophisticated. It was published by Macmillan and Company, in London, with illustrations by John Tenniel, on July 4, 1865. Carroll had suggested the date to commemorate the day, three years earlier, on which he had first extemporized his story.

The fourth and final version of *Alice* was a complete rewriting of the tale for very young children—children "from nought to five," as Carroll put it in his preface. This was brought out by Macmillan in 1889 with twenty of Tenniel's pictures enlarged and colored. (It is amusing to learn that Carroll rejected the first printing because the colors were too gaudy; the books were then offered to a New York publisher who refused them on the grounds that the colors were too dull.)

The 1965 Dover edition is a facsimile of the second manuscript, exactly as Carroll hand-lettered and illustrated it for Alice Liddell. In March 1885, Carroll obtained Alice's permission (she was then Mrs. Hargreaves) to allow Macmillan to issue the first facsimile of this manuscript. It went on sale on December 22 of the following year, in an edition of five thousand copies. On the last page of his original manuscript Carroll had pasted a small oval photograph of Alice, taken when she was seven, the age of Alice in the story. The upper part of this oval had separated the story's last two words. Because the photograph was removed from Macmillan's 1886 facsimile, the final line had to be relettered to close the gap.

Stuart Dodgson Collingwood, in his *Life and Letters of Lewis Carroll*, prints two letters from Carroll to Alice: one dated March 1, 1885, requesting her permission to publish the facsimile; the other, dated November 11, 1886, reporting his progress in preparing this edition. Both letters are so characteristic of Carroll, so revealing of his fondness for Alice Liddell, and so filled with interesting details about the history of the first facsimile, that they are worth quoting in full:

My dear Mrs. Hargreaves:

I fancy this will come to you almost like a voice from the dead, after so many years of silence, and yet those years have made no difference that I can perceive in *my* clearness of memory of the days when we *did* correspond. I am getting to feel what an old man's [Carroll was then fifty-three] failing memory is as to recent events and new friends (for instance, I made friends, only a few weeks ago, with a very nice little maid of about twelve, and had a walk with her—and now I can't recall either of her names!), but my mental picture is as vivid as ever of one who was, through so many years, my ideal child-friend. I have had scores of child-friends since your time, but they have been quite a different thing.

However, I did not begin this letter to say all *that*. What I want to ask is, Would you have any objection to the original MS. book of "Alice's Adventures" (which I suppose you still possess) being published in facsimile? The idea of doing so occurred to me only the other day. If, on consideration, you come to the conclusion that you would rather *not* have it done, there is an end of the matter. If, however, you give a favourable reply, I would be much obliged if you would lend it to me (registered post, I should think, would be safest) that I may consider the possibilities. I have not seen it for about twenty years, so am by no means sure that the illustrations may not prove to be so awfully bad that to reproduce them would be absurd.

There can be no doubt that I should incur the charge of gross egoism in publishing it. But I don't care for that in the least, knowing that I have no such motive; only I think, considering the extraordinary popularity the books have had (we have sold more than 120,000 of the two), there must be many who would like to see the original form.

Always your friend,
C. L. Dodgson.

My dear Mrs. Hargreaves:

Many thanks for your permission to insert "Hospitals" in the Preface to your book. I have had almost as many adventures in getting that unfortunate facsimile finished, *Above* ground, as your namesake had *Under* it!

First, the zincographer in London, recommended to me for photographing the book, page by page, and preparing the zinc-blocks, declined to undertake it unless I would entrust the book to *him*, which I entirely refused to do. I felt that it was only due to you, in return for your great kindness in lending so unique a book, to be scrupulous in not letting it be even *touched* by the workmen's hands. In vain I offered to come and reside in London with the book, and to attend daily in the studio, to place it in position to be photographed, and turn over the pages as required. He said that could not be done because "other authors' works were being photographed there, which must on no account be seen by the public." I undertook not to look at *anything* but my own book; but it was no use; we could not come to terms.

Then —— recommended me a certain

Mr. X—— , an excellent photographer, but in so small a way of business that I should have to *prepay* him, bit by bit, for the zinc-blocks; and *he* was willing to come to Oxford, and do it here. So it was all done in my studio, I remaining in waiting all the time, to turn over the pages.

But I daresay I have told you so much of the story already.

Mr. X —— did a first-rate set of negatives, and took them away with him to get the zinc-blocks made. These he delivered pretty regularly at first, and there seemed to be every prospect of getting the book out by Christmas, 1885.

On October 18, 1885, I sent your book to Mrs. Liddell, who had told me your sisters were going to visit you and would take it with them. I trust it reached you safely?

Soon after this—I having prepaid for the whole of the zinc-blocks—the supply suddenly ceased, while twenty-two pages were still due, and Mr. X —— disappeared!

My belief is that he was in hiding from his creditors. We sought him in vain. So things went on for months. At one time I thought of employing a detective to find him, but was assured that "all detectives are scoundrels." The alternative seemed to be to ask you to lend the book again, and get the missing pages re-photographed. But I was most unwilling to rob you of it again, and also afraid of the risk of loss of the book, if sent by post—for even "registered post" does not seem *absolutely* safe.

In April he called at Macmillan's and left *eight* blocks, and again vanished into obscurity.

This left us with fourteen pages (dotted up and down the book) still missing. I waited awhile longer, and then put the thing into the hands of a solicitor, who soon found the man, but could get nothing but promises from him. "You will never get the blocks," said the solicitor, "unless you frighten him by a summons before a magistrate." To this at last I unwillingly consented: the summons had to be taken out at —— (that is where this aggravating man is living), and this entailed two journeys from Eastbourne—one to get the summons (my *personal* presence being necessary), and the other to attend in court with the solicitor on the day fixed for hearing the case. The defendant didn't appear; so the magistrate said he would take the case in his absence. Then I had the new and exciting experience of being put into the witness-box, and sworn, and cross-examined by a rather savage magistrate's clerk, who seemed to think that, if he only bullied me enough, he would soon catch me out in a falsehood! I had to give the magistrate a little lecture on photo-zincography, and the poor man declared the case was so complicated he must adjourn it for another week. But this time, in order to secure the presence of our slippery defendant, he issued a warrant for his apprehension, and the constable had orders to take him into custody and lodge him in prison, the night before the day when the case was to come on. The news of *this* effectually frightened him, and he delivered up the fourteen negatives (he hadn't done the blocks) before the fatal day arrived. I was rejoiced to get them, even though it entailed the paying a second time for getting the fourteen blocks done, and withdrew the action.

The fourteen blocks were quickly done and put into the printer's hands; and all is going on smoothly at last: and I quite hope to have the book completed, and to be able to send you a very special copy (bound in white vellum, unless you would prefer some other style of binding) by the end of the month.

Believe me always,
Sincerely yours,
C. L. Dodgson.

The original manuscript of *Alice's Adventures Under Ground* was sold in April 1928 to A. S. W. Rosenbach, of Philadelphia, who resold it six months later to Eldridge R. Johnson, of Moorestown, New Jersey. According to Warren Weaver, who gives these details in his book *Alice in Many Tongues* (University of Wisconsin Press, 1964), after Johnson's death the manuscript was sold back to Rosenbach in 1946 for $50,000. Luther H. Evans, then Librarian of Congress, felt strongly that the manuscript belonged in England. He persuaded Rosenbach to sell it again (after raising $50,000 from various persons), and in 1948 Evans gave it to the British Museum, where it remains today. Weaver quotes the Archbishop of Canterbury, who accepted the gift for the museum, as saying that its return to England was "an unsullied and innocent act in a distracted and sinful world."

Several facsimiles have been published in the United States. The present Dover facsimile was prepared with the cooperation of the British Museum, directly from photographs of the original manuscript. Carroll's title and dedication pages are reproduced in their original color on the front and back covers of this edition, though the colors he gave to initial letters of chapter headings could not be shown. The book is accurate in every respect, including the photograph of Alice, still attached to the final page of the original.

In addition to Carroll's manuscript book, this edition contains the following material, reproduced from the Macmillan 1886 facsimile: Carroll's preface, his "Easter Greeting," his poem called "Christmas Greetings," and the book's two pages of advertising. These, with the exception of the preface, appeared at the end of the Macmillan volume. The design facing the Dover title page is that which appeared on the binding of the 1886 Macmillan facsimile.

It is difficult today to read Carroll's preface and "Easter Greeting" without embarrassment. "No shadow of sin," indeed! One wonders how he would have explained a battle between two rival siblings over, say, who gets the last cookie in the jar. "Innocent laughter" and "merry voices" as youngsters "roll among the hay"? Carroll has a point, but it is so plastered over with what H. L. Mencken liked to call "pious piffle," and with rationalizations for his abnormal devotion to little girls, that I suspect even the most devout Victorians shuddered a bit as they read.

Many of Carroll's changes, when he revised this story for its first book publication, consisted of the addition of entirely new characters, new episodes, and new poems. The Mad Tea Party and the Pig and Pepper chapters were added. The trial of the Knave of Hearts was enlarged from the less than two pages it occupies here to two new chapters. Neither the

Cheshire Cat nor the Ugly Duchess is in the original, though there is a briefly mentioned Marchioness who later turns out to be the Queen. That famous shaped poem, the Mouse's Tail, is completely different here (p. 28). The episode on pages 26 and 27, in which the Dodo leads everyone to a cottage to get dry, recalls an occasion in 1862 when Dodgson and Duckworth (the Dodo and the Duck) and the three Liddell girls had actually been caught in a downpour during another boating trip. Carroll had led the group (it also included two of Carroll's sisters and his aunt Lucy: the "several other curious creatures" mentioned on p. 22) to a friend's home where they dried their clothes. Carroll assumed this incident would be of little interest outside the small circle of those involved, so he took it out and substituted the story of the Caucus Race.

Carroll also later removed, from pages 14 and 15, the names of Gertrude and Florence. They were probably friends or relatives of Alice, and since the references to Florence were unkind, he changed the names to Ada and Mabel. Knowledgeable readers will spot other details in which the manuscript differs from the familiar version: the White Rabbit's nosegay (p. 13) that later became a fan, the Caterpillar's distinction between the mushroom's top and stalk (p. 61) that later became a distinction between right and left sides, the ostrich (p. 76) that became a flamingo, and so on.

Tenniel saw Carroll's drawings before he made his own sketches. There are obvious similarities here and there, but some such resemblances would have been hard to avoid. Carroll was not much of an artist, as he himself was fully aware, but his illustrations are nonetheless fascinating. They show, more accurately than Tenniel's drawings, how Carroll imagined his curious characters. And there are amusing details, such as the kissing courtiers on page 75 and some not consciously intended aspects of special interest to Freudian symbol searchers, that call for more than a casual inspection of the pictures.

Postscript

In 1977 Morton Cohen, the eminent Carrollian biographer and critic, was in the British Museum examining the original manuscript of *Alice's Adventures Under Ground.* He made a surprising discovery. Underneath the oval photograph of Alice's face that Carroll had pasted at the end of the manuscript was a drawing Carroll had made of the face! The museum now has the photo on a hinge so it can be lifted up to reveal Carroll's sketch.

Cohen tells the story of how he found the drawing in the *New York Times Book Review* (October 19, 1977), and in his short article "Alice Under Ground," *Jabberwocky* (Autumn 1978).

36.

THE TWO ALICE BOOKS

This reprints my introduction to a Signet paperback edition of *Alice's Adventures in Wonderland* and *Through the Looking-Glass* (1960).

Although many adults dislike fantasy, preferring fiction about the real world, it is surprising how many great literary works are fantasies. One thinks of Homer's *Iliad* and *Odyssey*, Virgil's *Aenead*, Dante's *Divine Comedy*, Goethe's *Faust*, Shakespeare's *Tempest* and *Midsummer Night's Dream*, Milton's *Paradise Lost*, and scores of fantasy novels that have outlasted myriads of once admired works of realism.

Among books for children almost all the classics are fantasies: P. L. Travers's *Mary Poppins*, Barrie's *Peter Pan*, French fairy tales by Perrault, German fairy tales by Grimm, Danish tales by Andersen, Italy's *Pinocchio*, Kipling's *Just-So* stories, C. S. Lewis's *Narnia*, the Oz books of L. Frank Baum, and of course Lewis Carroll's two books about Alice. At the moment, J. K. Rowling's incredibly popular books about Harry Potter, boy wizard, are fantasies. Whether they will become classics remains to be seen.

It's hard to understand, but some adults, including a few peculiar psychologists, think fantasy is bad for children. G. K. Chesterton considered this belief close to mortal sin. His marvelous essay "The Dragon's Grandmother" (you'll find it in his book *Tremendous Trifles*) is, in my opinion, the best defense of juvenile fantasy ever written.

"I met a man the other day," G. K. opens his essay, "who did not believe in fairy tales. I do not mean that he did not believe in the incidents narrated in them. . . . The man I speak of disbelieved in fairy tales in an even more amazing and perverted sense. He actually thought that fairy tales ought not to be told to children."

Chesterton recalls his efforts to wade through some modern novels about the actual world. When this became tiresome he saw a copy of Grimm on the table and gave "a cry of indecent joy. Here at least, here at last, one could find a little common sense. I opened the book and my eyes fell on those splendid and satisfying words, 'The Dragon's Grandmother.'"

At that moment, the monster who hated fairy tales entered the room. Here is what Chesterton said to him:

> It is far easier to believe in a million fairy tales than to believe in one man who does not like fairy tales. I would rather kiss Grimm instead of a Bible and swear to all his stories as if they were thirty-nine articles than say seriously and out of my heart that there can be such a man as you; that you are not some temptation of the devil or some delusion from the void. Look at these plain, homely, practical words. "The Dragon's Grandmother," that is all right; that is rational almost to the verge of rationalism. If there was a dragon, he had a grandmother. But you—you had no grandmother! If you had known one, she would have taught you to love fairy tales.
>
> It seemed to me that he did not follow me with sufficient delicacy, so I moderated my tone. "Can you not see," I said, "that fairy tales in their essence are quite solid and straightforward; but that this everlasting fiction about modern life is in its nature essentially incredible? Folk-lore means that the soul is sane, but that the universe is wild and full of marvels. Realism means that the world is dull and full of routine, but that the soul is sick and screaming. The problem of the fairy tale is—what will a healthy man do with a fantastic world? The problem of the modern novel is—what will a madman do with a dull world? In the fairy tales the cosmos goes mad; but the hero does not go mad. In the modern novels the hero is mad before the book begins, and suffers from the harsh steadiness and cruel sanity of the cosmos.

"The cosmos goes mad; but the hero does not go mad." It's an accurate description of Alice as she strolls through the mad world of Wonderland and the equally mad world behind the looking-glass. It is an accurate description of Dorothy as she walks the Yellow Brick Road in the mad world of Oz.

There really are curious persons so down on fantasy that they find no pleaseure in Carroll's Alice books. Consider what H. L. Mencken says about them in his autobiographical *Happy Days*:

> I was a grown man, and far gone in sin, before I ever brought myself to tackle "Alice in Wonderland," and even then I made some big skips and wondered sadly how and why such feeble jocosity had got so high a reputation. I am willing to grant that it must be a masterpiece, as my betters allege—but not to *my* taste, not for *me*.

Lewis Carroll, whose real name was Charles Lutwidge Dodgson, was a shy bachelor who taught elementary mathematics at Oxford University's Christ Church, and who had a passion for photography when that art was in its infancy. He loved attractive little girls much more than boys. He was especially fond of the real Alice, Alice Liddell, daughter of the dean of Christ Church. Her last name rhymes with "fiddle." Carroll puns three times on the name's closeness to "little" in his prefatory poem to the first Alice book, and again in chapter 7 where he writes about the three little girls (Alice had two sisters) who lived at the bottom of a well. There is convincing evidence that Carroll was romantically in love with Alice.

Morton Cohen, in his splendid biography of Carroll, speculates that Carroll may actually have approached Mr. and Mrs. Liddell with a suggestion that he wished to marry Alice someday. At any rate, Mrs. Liddell suddenly decided that Carroll should stop seeing Alice, and she burned all his letters to her. The page in Carroll's diary for the time when this occurred has been torn from the diary and presumably destroyed. It is widely believed that in the second Alice book, Carroll intended the White Knight to be a caricature of himself. When he waves good-bye to Alice, as she moves to the final square of the chessboard to become a queen, the scene represents his final sad parting with the only child-friend he truly loved.

Among Carroll's superb photographs of famous persons and young children were hundreds of pictures of unclad little girls. After gossip implied there was something unhealthy in such picture taking, Carroll destroyed all his nude negatives. The four that survived are reproduced in Morton Cohen's beautiful volume, *Reflections in a Looking Glass: A Centennial Celebration of Lewis Carroll, Photographer*.

There is no question that Carroll's fondness for young girls, and photographing them without clothes, was in his mind a completely innocent admiration of their charms. He was a devout Anglican, an ordained deacon who often preached in nearby churches. His religious opinions were strictly orthodox except for a heretical refusal to believe God would eternally punish anyone, including Satan. Entries in his dairy constantly repeat a yearning to obey God's will and to be forgiven for sins that are never specified.

Lewis Carroll was born in 1832, in Cheshire, England, the third of eleven children of the Reverend Charles Dodgson, an Anglican priest. Although Carroll's major hobby was photography, he was also fond of showing magic tricks to child friends and taking them to see conjuring shows. He liked to form a handkerchief into a mouse, then make it jump from his hand when he stroked it. He would fold from paper what he called "paper pistols" that made a loud bang when swung through the air. He was an avid patron of the theater, though frequently offended by stage profanity and scenes he considered sacrilegious or licentious. The famous actress Ellen Terry was a lifelong friend.

The first Alice book had its origin in a story Carroll told Alice and her two sisters when they were on a rowing trip up the Isis, a branch of the Thames. Alice begged him to write it down for her. This he did, calling it *Alice's Adventures Under Ground*. Carroll himself illustrated the tale, though his artistic ability, as he recognized, was poor. He presented a hand-lettered copy to Alice, and it was later published as a small book. The story differs in many interesting ways from the final printed version, and it lacks such episodes as the Pig and Pepper chapter and the Mad Tea Party. If you are curious to read it, Dover has a paperback reprint.

Both Alice books reflect Carroll's interest in games. Among his dozens of privately printed pamphlets, many describe games he himself invented. They include variations of croquet, games played with pieces on a checkerboard, and a game he called "doublets" in which one tried in the fewest possible steps to change one word to another by altering one letter at a time, at each stop forming a common word. In the first Alice book, playing cards provide the roles of courtiers, queens, kings, and laborers. In the sequel, cards are replaced by the kings, queens, and knights of chess. Each book is rich in wordplay, nonsense verse, mathematical whimsy, and logical paradoxes.

The Alice books swarm with sly implications that Victorian readers would grasp at once, but today have to be explained by footnotes such as you will find in the 2000 edition of my *Annotated Alice*. I will give you only one amusing instance. When Humpty Dumpty shakes hands with Alice he extends only one finger. In Victorian days, when someone shook hands with a person of inferior social status, it was customary to extend only two fingers. Humpty, in the great pride that went before his fall, carries this ugly custom to its extreme.

Tenniel's illustrations also have subtle features that are easily missed. Look carefully at the picture of Humpty seated on the wall. A cross section of the wall on the right reveals that the wall has a narrow ridge, like an inverted V. This of course renders the egg's seat extremely precarious. In the picture of the Queen of Hearts asking Alice "What's your name, child?" see if you can find the white rabbit. After Alice has gone through the looking-glass, note the grin on the back of the clock and the gargoyle face below the mantel. Tenniel slyly shaded the nose of the Jack of Hearts to suggest that he was a heavy drinker. For much more about Tenniel and his art see *The Tenniel Illustrations to the "Alice" Books*, a delightful study by Michael Hancher (1985).

Alice's Adventures in Wonderland was first published by London's Macmillan in 1865. The covers are red cloth, with gold lettering on the spine. On the front cover is a picture of Alice holding a pig; on the back a picture of the Cheshire Cat. The print run was two thousand. Because Tenniel strongly objected to the quality of the printing of his pictures, Macmillan was forced at considerable expense to make a new printing. Copies of the first run were sold to an American publisher. The book was such a success that by 1899 there had been twenty-six new printings. Carroll continued to correct printer's errors and make small changes in the text until the eighty-nine thousandth was issued in 1897, a year before his death.

Macmillan published *Through the Looking-Glass* in 1872. It, too, was bound in red cloth, with the Red Queen on the front cover and the White

Queen on the back. It sold even faster than the first Alice book. Fifteen thousand copies were bought in the first seven weeks.

After Carroll's death, at age sixty-five, new editions of his Alice books began to appear with illustrations by other artists. In 1901, Harper and Brothers published in America an edition of the first Alice book with picture by the American artist Peter Newell. The following year it issued Newell's illustrated edition of the second Alice book. His eighty full-page plates are reproduced in my *More Annotated Alice* (1990). Since then scores of artists, including the famous illustrator Arthur Rackham, have tried their hand on the Alice books. *The Illustrators of Alice* (1972), edited by Graham Overden, is devoted to the history of illustrated editions of Alice, with many handsome reproductions of their art. Some recent artists take extreme liberties. For example, Barry Moser's art for the second Alice book gives Humpty the face of Richard Nixon!

So many translations of Alice into other languages have been made that mathematician and Carroll collector Warren Weaver wrote an entire book about them, *Alice in Many Tongues* (1964). So many plays and musicals were based on Alice that Charles C. Lovett covers their history in *Alice on Stage* (1990). A checklist of motion pictures about Alice, by David Schaefer, will be found at the back of the Norton edition of my *Annotated Alice*.

Many biographies of Carroll have been written, and several book collections made of his correspondence. A two-volume expurgated edition of Carroll's diary, edited by Roger Lancelyn Green, appeared in 1954. An uncut edition of the diary, edited by Edward Wakeling, is now being issued in many volumes. Alice Liddell has her own biography, *The Real Alice* (1981), by Anne Clarke.

Aside from his books about mathematics and logic, and collections of his poetry, Carroll wrote two other fantasies. A long novel about two fairy children was published in two volumes: *Sylvie and Bruno* (1889) and *Sylvie and Bruno Concluded* (1893). His great nonsense balled *The Hunting of the Snark*, illustrated by Henry Holiday, appeared in 1876. You'll find its history, along with numerous footnotes, in my *Annotated Hunting of the Snark* (2006). The ballad tells of a crew of bizarre adventurers who set sail to find an elusive monster called the Snark. It also was illustrated by Peter Newell, as well as by many other artists since.

Carroll's long comic poem *Phantasmagoria*, about a friendly ghost, was recently published by Prometheus Books. Several books deal with Carroll's writings about puzzles and recreational mathematics, including my *Universe in a Handkerchief* (1996). For a detailed and annotated bibliography of all of Carroll's writings, the classic reference is *The Lewis Carroll Handbook*, by Sidney Herbert Williams and Falconer Madan, revised and updated by Roger Green in 1962. Other bibliographies, essay collections, and critical studies of Carroll's work are far too numerous to mention.

The Lewis Carroll Society of North America publishes a news bulletin entitled the *Knight Letter*, and sponsors two annual conventions. In England, an older Lewis Carroll Society, founded in 1969, issues a scholarly periodical, the *Carrollian* (formerly *Jabberwocky*), edited by Anne Clarke Amor, and a newsletter called *Bandersnatch*. A Canadian Carroll Society publishes *White Rabbit Tales*, its newsletter, and there is a recently organized Carroll Society of Japan. Carroll has an extensive following in Japan, where some sixty editions of the Alice books are currently in print!

You won't find the Alice books reprinted in Mortimer Adler's set of *The Great Books of the Western World*, but I venture to state the following: It is permissible today to consider a person not educated if he or she has not read, or is not at least familiar with, *Das Capital*, or books by Hegel and Freud, or indeed more than half the volumes in Adler's series. On the other hand, I would not consider a person educated who has never read Carroll's Alice books. Among the great characters of imaginative literature, Alice has become as immortal as Don Quixote, Huckleberry Finn, Captain Ahab, Sherlock Holmes, and Dorothy Gale of Kansas. If you are not yet acquainted with Alice's adventures in Wonderland and behind the mirror, read on and enjoy!